Communications in Computer and Information Science 1231

Commenced Publication in 2007
Founding and Former Series Editors:
Simone Diniz Junqueira Barbosa, Phoebe Chen, Alfredo Cuzzocrea,
Xiaoyong Du, Orhun Kara, Ting Liu, Krishna M. Sivalingam,
Dominik Ślęzak, Takashi Washio, Xiaokang Yang, and Junsong Yuan

More information about this series at http://www.springer.com/series/7899

Piotr Gaj · Wojciech Gumiński ·
Andrzej Kwiecień (Eds.)

Computer Networks

27th International Conference, CN 2020
Gdańsk, Poland, June 23–24, 2020
Proceedings

Editors
Piotr Gaj (iD)
Silesian University of Technology
Gliwice, Poland

Wojciech Gumiński (iD)
Gdańsk University of Technology
Gdańsk, Poland

Andrzej Kwiecień (iD)
Silesian University of Technology
Gliwice, Poland

ISSN 1865-0929   ISSN 1865-0937 (electronic)
Communications in Computer and Information Science
ISBN 978-3-030-50718-3   ISBN 978-3-030-50719-0 (eBook)
https://doi.org/10.1007/978-3-030-50719-0

This Springer imprint is published by the registered company Springer Nature Switzerland AG
The registered company address is: Gewerbestrasse 11, 6330 Cham, Switzerland

Preface

Communication is one of the most important parts of the modern world. The field of computer networks constantly delivers new technologies that meet expectations of all services required by various applications, from many years. They are like digital bloodstream of computer systems, highly necessary but invisible unless they are damaged. Most of devices and whole systems are dependent on computer networks services. They become useless or their functionalities would be significantly diminished if the computer networks stopped working. Because the majority of digital facilities are a part of networked systems, the communication technologies are one the most important part of them ensuring their operation. Try to imagine, what personal and professional activities would be in time of COVID-19 lockdown, if we did not have the significant development of networks communication in recent years. Therefore, the prospects in this domain bring inevitable development, and events founded to exchange and disseminate knowledge in this area are important.

It is indispensable to have an in-depth knowledge of how to manage, model, and design networked systems. To the high dynamics and the multiplicity of emerging technologies, it is necessary to constantly expand and exchange knowledge and gain experiences in this field. Conferences are the kind of meetings where independent points of view are presented and where experts, researchers, and users can exchange ideas. This book contains the top proceedings of one such event.

The Computer Networks (CN) conference was established at the Faculty of Automatic Control, Electronics and Computer Science of Silesian University of Technology in Gliwice 25 years ago. Professor Andrzej Grzywak was the main initiator of this event. The 27th edition of the conference took place in 2020, and we hope that more editions will be held in the future. With the increasing popularity of this topic, we are facing the difficult task of dealing with numerous article submissions with twice as much strength.

The innovative solutions and proposals submitted to CN indicates that significant, relative scientific research is discussed. Every year, the number of publications is growing and the scientific quality of works is getting more advanced. The scientific level of the presented works is very high, and as a consequence the reviewing process is very demanding and difficult for the authors. It consists of three independent opinions of well-known scientists from around the world. Computer networks are still the main solution which allows nodes to share resources. This is very important from many points of view, including industrial communication.

Since 2009, conference proceedings have been published by Springer in the CCIS series. All conference issues are indexed by Web of Science and Scopus each year. It is also worth mentioning that the conference has co-sponsors and co-organizers, which include the Computer Networks and Distributed Systems Section of the Informatics Committee of the Polish Academy of Sciences (PAN) as well as IEEE Poland and the International Network for Engineering Education and Research (iNEER).

27th International Science Conference on Computer Networks (CN 2020)

The CN conference has been visible for more than a quarter of a century, giving researchers a chance to meet each other, make new connections, start cooperation, discuss on bothering problems, as well as disseminate their research results. The essential contents presented during the conference are published in significant and well recognized series of proceedings. Over the past 26 years of the conference's history, all important topics related to computer networks area have been discussed at the conference and major breakthroughs in this area have been deliberated. Many collaborative relationships were established, both in local and international scope. Thus, we believe that the event has had a significant contribution to the global pool of achievements in this domain.

Computer networks are still the only communication means for digital systems of all kinds. Thus, recent research and innovative applications are very important for current industrial and social activities. We also expect a high need for this in the future. Computer networks and internal complexity of their operations are usually not shown to their user. But without modern solutions and developments many popular and spectacular amenities of everyday life would be unavailable.

This year, the CN conference faced the non-precedent situation related to both pandemic state and internal regulations changes in Poland. It caused a reduced number of submissions as well as the withdrawal of some works. Despite this, the organizers ran the conference with the best topics among submitted and by including here the best articles related to them. For the current edition, nearly 50 papers were submitted. To maintain the high quality of the CCIS publication, only 34 were selected for further consideration, and 14 among them were carefully selected for publication in this proceedings. Each paper was reviewed by three independent reviewers in a double-blind process. The Technical Program Committee of CN 2020 consisted of 136 scientists from 25 countries and 5 continents. This book collects the research work of scientists from notable research centers. It includes stimulating studies of the wide spectrum of both science and practical-oriented issues regarding the computer networks and communication domain that may interest a wide readership. The content is divided in three parts.

- Computer Networks

   This section contains seven papers. All of them refer to the general domain of computer networks and communications problems.

The first paper is delivered by two research entities: the Institute of Analysis and Scientific Computing, TU Wien, Austria and the Institute of Computer Science and Information Technologies, Lvov Polytechnic National University, Ukraine. It refers to the issue of finding the proper network topology regarding the criteria of bandwidth utilization. The Gomory-Hu algorithm is considered which is modified with taking into account deficient channels. Network topology optimization made by the proposed algorithm guarantees the transmission of the maximum input stream. The presented idea, although based on classical Gomory-Hu algorithm, seems to be very promising. The authors provide also an example showing that the result obtained with the modified algorithm is correct.

The second paper is presented by the representatives of the University of Houston, USA. The authors ponder communication networks used in space. The space technologies become continuously more important last years and communication on this matter is highly relevant to develop the whole branch. Authors examined the segmentation process used by Licklider Transmission Protocol to determine the role of the segment length in the context of transmission delays. They propose the model of the protocol that allows estimating the variable length of data chunk instead of fixed one. This innovative approach is possible because authors provide a relation between bit error rate of a channel and the optimal segment length instead of common practice of using the maximum payload of underlying protocol.

The third paper is prepared by scientists from the Technical University of Liberec, Czech Republic, and refers to the NAT64 and DNS64 mechanisms. The authors specify the problems related to the DNS use in modern network services, especially problems related to default usage of third party DNS resolvers together with the most deployed detection method described by RFC7050. These issues could become real, as the method is not compatible with DoH resolvers. This could even lead to problems which would prevent Internet service providers from disabling IPv4 in their network. Authors suggest how to solve the issues related to RFC7050 and show a possible way how to move information about both NAT64 and DNS64 from local view of top-level domain to operator's global zone, with keeping the security.

The authors of the next paper come from two Polish universities. The first author is from the University of Zielona Góra and the second from Military University of Technology. They touch a brand new technology, namely networks that transmit quantum information. Just like in regular networking one of the most important processes in quantum circuits is packet switching. The authors describe an implementation of a router for a four-qutrit quantum circuit. In general, the quantum router is a solution, working on qubits, and provides spin interactions between quantum units of information. Authors show that the joining of such routers allows for the building of structures which are able to transfer a quantum state to the defined node in a quantum network, achieving high accuracy of information transfer.

The next team is from Silesian University of Technology, Poland, and in their article authors analyze the state of the art in relation to the development of supercomputers and the important usage of related network technologies. They present the trends available in the domain of high-performance computers. The analysis is focused on system architecture, processors, computing accelerators, energy efficiency, and interconnection ability as well. The authors show that a significant impact on the supercomputer's development depends on many, various elements but the development of new topologies and technologies designed for connecting system nodes is absolutely pivotal.

The next paper has been produced in Germany by researchers from Technical University of Dresden. The aim of the paper is reliable delivery of crowd monitoring data. The authors discuss the dynamic switching between infrastructure network and peer-to-peer communication in a case when the connectivity is lost. The availability of such services is important from the emergency point of view of big events. The authors tested the P2P connection during an experiment they made at a real annual fair on the university campus.

The last paper in this section refers to 5G cellular networks and its author comes from Jagiellonian University, Poland. The development of this technology raises many unreasonable emotions these days, as new technologies usually do, but it is inevitable and finally will bring a positive impact to our networked society. The delivery of a small payload in a short time is requested in 5G and is possible by achieving ultra-reliable and low-latency communication. This is one of the major challenges in this kind of communication. In this paper, the authors provide some important definitions and present a method for reliability enhancement of such type of traffic. They consider the maximization of the reliability enhancement as an optimization problem and they make some relevant simulations to obtain an optimal resource allocation policy. As a result, they achieve significant performance differences between standard methods and the studied one.

– Cybersecurity and Quality of Service
  This section contains three papers related to networks security, reliability, and quality of service issues.

This first paper is made by author from the University of Ostrava, Czech Republic. The content of the paper ruminates the well-known and common problem of unwanted emails and anti-spam systems which block one from receiving them. The authors present the interconnection between two significant layers of multi-layer spam detection systems. Such communication is usually a weak point in mutual collaboration between many SMTP servers. Thus, the construction of the feedback interconnection between message content check and greylisting layers is a key, and authors propose an easy way how spam detection can be improved by this. It seems that the proposed method can improve the system's effectiveness because the obtained results prove that the number of detected spam messages is higher in comparison to the other well-known methods. The method is not related to the given IP version and is not connected with the particular implementation so it can be adopted to any multi-level spam detection system.

The authors of the second article come from Otto-Friedrich-University Bamberg, Germany, and V. A. Trapeznikov Institute of Control Sciences, Russian Academy of

Sciences, Russia. They consider an analysis of transport reliability in fog computing approach. The network function virtualization paradigm in an IoT scenario together with a software-defined networks stack and multipath communication between its clients and servers are used. The authors analyze the reliability of the redundant transport system. The used communication channels are error-prone simulated by random failures described by general Markov-modulated Poisson processes. They found that the steady-state distribution of the restoration model can be effectively calculated by a semi convergent iterative aggregation-disaggregation method for block matrices. As the result of the presented analysis authors compute the associated reliability function and hazard rate. The obtained effects seem to be useful for future generations of these kinds of networks.

The authors of the last paper in this section are from Gdańsk University of Technology, Poland. Both are from the Faculty of Electronics, Telecommunications and Informatics. They propose a new approach to network bandwidth distribution which can ensure so-called fairness to end-users. In some services where the high competition exists between users, e.g., multimedia transfer, the fairness assurance in the assignment of limited resources to a potentially large set of users competing for them is highly requested. The authors define fairness in terms of quality of experience for satisfied users and quality of service for unsatisfied users. Such an algorithm of fair bandwidth distribution can be one of the most desired supports for service providers, because its aim is minimizing the number of end-user service terminations. The proposed algorithm works much better than others commonly used, and as the result the number of resignations was almost four times lower than that of other algorithms.

– Queueing Theory and Queuing Networks
  This section contains four papers. The chapter refers to the theory of queues and queueing network models. In such models the time characteristics of all tasks at each network node is given by the response time of a queueing network.

Authors of the first paper in this section are from Petrozavodsk State University and the Institute of Applied Mathematical Research of the Karelian research centre of RAS, Russia. They talk about the implementation of failure rate functions to compare queueing processes for exponential, Pareto, and exponential-Pareto mixture distribution of service times. They use the failure rate and stochastic comparison techniques together with coupling of random variables to establish some monotonicity properties of the model. The obtained results can be useful for the estimation of the performance measures of a wide class of queueing systems.

The second paper comes from Poland. The authors are from the Institute of Computer Science, Cardinal Stefan Wyszyński University in Warsaw and from the Institute of Information Technology, Warsaw University of Life Sciences – SGGW. The authors report theoretical investigations on modified M/G/1/inf queuing model. They study this model with non-homogeneous customers, an unreliable server and a service time distribution which depends on a volume characteristic of the jobs. They extend their previous work and try to follow a purely transform-oriented analysis approach for M/G/1 models. The results show that the method of an additional event can be used in the case of complicated models. The queueing model may be relevant in virtualized computer systems.

The third paper is a result of research from USA, the University of California San Diego and California State University Northridge. The paper presents an infinite-server queue model with transient analysis and nonhomogeneous arrival processes. The paper is built on the previous defined mathematical models and presents a detailed analysis of the presented model. The authors obtained the basic differential equations for joint probability generating functions for a number of busy servers and served customers for transient and stationary random environments. The results seem to be suitable for network performance evaluation, as well as for designing the optimal strategies for managing resources of various networked systems and subsystems where the considered model can be used.

The last paper is a result of international cooperation. The authors are from, Shri Mata Vaishno Devi University, India, and the Polish Academy of Science as well as Silesian University of Technology, Poland. The authors consider cloud services which are provided by virtual machines. They presented a simple queuing model for processing tasks in computational clouds. The research is based on transient analysis of the performance parameters. The numerical examples are presented to illustrate its utility by considering the effects of reneging and feedback on the queueing delay, probability of task rejection, and the probability of immediate service.

On behalf of the Program and Organizing Committee of the CN conference, we would like to express our gratitude to all authors for sharing their research results and for their assistance in producing this volume, which we believe is a reliable reference in the computer networks domain.

We also want to thank the members of the Technical Program Committee and all reviewers for their involvement and participation in the reviewing process.

If you would like to help us make the CN conference better, please send us your opinions and suggestions at cn@polsl.pl.

May 2020   Piotr Gaj
                                                                Wojciech Gumiński
                                                                Andrzej Kwiecień

Organization

CN 2020 was organized by the Faculty of Automatic Control, Electronics and Computer Science, Silesian University of Technology (SUT) and supported by the Committee on Informatics of the Polish Academy of Sciences (PAN), the Section of Computer Networks and Distributed Systems in technical co-operation with the IEEE, and consulting support of the iNEER organization.

The official co-organizer was the Faculty of Electronics, Telecommunications and Informatics (ETI) of the Gdańsk University of Technology (GUT) in Gdańsk, Poland.

Executive Committee

The Executive Committee are from the Silesian University of Technology (SUT) and the Gdańsk University of Technology (GUT), Poland.

The SUT Team

Honorary Member

Halina Węgrzyn   Silesian University of Technology, Poland

Conference Manager

Piotr Gaj   Silesian University of Technology, Poland

Deputy Manager

Jacek Stój   Silesian University of Technology, Poland

Technical Support

Aleksander Cisek   Silesian University of Technology, Poland

Technical Support

Ireneusz Smołka   Silesian University of Technology, Poland

Office

Małgorzata Gładysz   Silesian University of Technology, Poland

Web Support

Piotr Kuźniacki   Silesian University of Technology, Poland

The GUT Team

Deputy Manager

Jacek Rak   Gdańsk University of Technology, Poland

Technical Volume Editor

Wojciech Gumiński   Gdańsk University of Technology, Poland

Co-ordinators

ETI Co-ordinator

Zenon Filipiak   Gdańsk University of Technology, Poland

PAN Co-ordinator

Tadeusz Czachórski   Gdańsk University of Technology, Poland

IEEE PS Co-ordinator

Marek Jasiński   Gdańsk University of Technology, Poland

iNEER Co-ordinator

Win Aung   Gdańsk University of Technology, Poland

Program Committee

Program Chair

Andrzej Kwiecień   Silesian University of Technology, Poland

Honorary Members

Win Aung   iNEER, USA
Joanna Polańska   Silesian University of Technology, Poland
Bogdan M. Wilamowski   Auburn University, USA
Jerzy Wtorek   Gdańsk University of Technology, Poland

Technical Program Committee

Davide Adami   University of Pisa, Italy
Tülin Atmaca   Institut National de Télécommunication, France
Rajiv Bagai   Wichita State University, USA
Sebastian Bala   University of Opole, Poland
Jiří Balej   Mendel University in Brno, Czech Republic
Alexander Balinsky   Cardiff University, UK

Anna Kamińska-Chuchmała	Wrocław University of Science and Technology, Poland
Jerzy Klamka	IITiS Polish Academy of Sciences, Poland
Wojciech Kmiecik	Wrocław University of Science and Technology, Poland
Grzegorz Kołaczek	Wrocław University of Science and Technology, Poland
Ivan Kotuliak	Slovak University of Technology in Bratislava, Slovakia
Zbigniew Kotulski	Warsaw University of Technology, Poland
Demetres D. Kouvatsos	University of Bradford, UK
Stanisław Kozielski	Silesian University of Technology, Poland
Henryk Krawczyk	Gdańsk University of Technology, Poland
Udo R. Krieger	Otto-Friedrich-University Bamberg, Germany
Michał Kucharzak	Wrocław University of Technology, Poland
Rakesh Kumar	Shri Mata Vaishno Devi University, India
Mirosław Kurkowski	Police Academy in Szczytno, Poland
Andrzej Kwiecień	Silesian University of Technology, Poland
Piotr Lech	West-Pomeranian University of Technology, Poland
Piotr Lechowicz	Wrocław University of Technology, Poland
Ricardo Lent	University of Houston, USA
Jerry Chun-Wei Lin	Western Norway University of Applied Sciences, Norway
Zbigniew Lipiński	University of Opole, Poland
Wolfgang Mahnke	ascolab GmbH, Germany
Aleksander Malinowski	Bradley University, USA
Marcin Markowski	Wrocław University of Science and Technology, Poland
Przemysław Mazurek	West-Pomeranian University of Technology, Poland
Agathe Merceron	Beuth University of Applied Sciences, Germany
Jarosław Miszczak	IITiS Polish Academy of Sciences, Poland
Vladimir Mityushev	Pedagogical University of Cracow, Poland
Jolanta Mizera-Pietraszko	University of Opole, Poland
Evsey Morozov	Petrozavodsk State University, Russia
Włodzimierz Mosorow	Lodz University of Technology, Poland
Sasa Mrdovic	University of Sarajevo, Bosnia and Herzegovina
Mateusz Muchacki	Pedagogical University of Cracow, Poland
Gianfranco Nencioni	University of Stavanger, Norway
Sema F. Oktug	Istanbul Technical University, Turkey
Remigiusz Olejnik	West Pomeranian University of Technology, Poland
Michele Pagano	University of Pisa, Italy
Nihal Pekergin	University Paris-Est Créteil, France
Maciej Piechowiak	University of Kazimierz Wielki in Bydgoszcz, Poland
Piotr Pikiewicz	College of Business in Dąbrowa Górnicza, Poland
Jacek Piskorowski	West Pomeranian University of Technology, Poland
Bolesław Pochopień	Silesian University of Technology, Poland

Oksana Pomorova	University of Lodz, Poland
Sławomir Przyłucki	Lublin University of Technology, Poland
Jacek Rak	Gdańsk University of Technology, Poland
Tomasz Rak	Rzeszow University of Technology, Poland
Stefan Rass	University of Klagenfurt, Austria
Stefano Rovetta	University of Genoa, Italy
Przemysław Ryba	Wrocław University of Science and Technology, Poland
Vladimir Rykov	Russian State Oil and Gas University, Russia
Wojciech Rząsa	Rzeszow University of Technology, Poland
Dariusz Rzońca	Rzeszow University of Technology, Poland
Alexander Schill	TU Dresden, Germany
Olga Siedlecka-Lamch	Czestochowa University of Technology, Poland
Artur Sierszeń	Lodz University of Technology, Poland
Mirosław Skrzewski	Silesian University of Technology, Poland
Adam Słowik	Koszalin University of Technology, Poland
Pavel Smolka	University of Ostrava, Poland
Tomas Sochor	University of Ostrava, Czech Republic
Maciej Stasiak	Poznań University of Technology, Poland
Ioannis Stylios	University of the Aegean, Greece
Grażyna Suchacka	University of Opole, Poland
Wojciech Sułek	Silesian University of Technology, Poland
Zbigniew Suski	Military University of Technology, Poland
Bin Tang	California State University, USA
Kerry-Lynn Thomson	Nelson Mandela Metropolitan University, South Africa
Oleg Tikhonenko	Cardinal Stefan Wyszynski University, Poland
Ewaryst Tkacz	Silesian University of Technology, Poland
Homero Toral Cruz	University of Quintana Roo, Mexico
Mauro Tropea	University of Calabria, Italy
Leszek Trybus	Rzeszów University of Technology, Poland
Kurt Tutschku	Blekinge Institute of Technology, Sweden
Adriano Valenzano	National Research Council of Italy, Italy
Bane Vasic	University of Arizona, USA
Miroslaw Voznak	VSB-Technical University of Ostrava, Czech Republic
Sylwester Warecki	Broadcom Ltd., USA
Jan Werewka	College of Economics and Computer Science, Poland
Tadeusz Wieczorek	Silesian University of Technology, Poland
Lukasz Wisniewski	OWL University of Applied Sciences, Germany
Przemysław Włodarski	West Pomeranian University of Technology, Poland
Józef Woźniak	Gdańsk University of Technology, Poland
Hao Yu	Xilinx, USA
Grzegorz Zaręba	Accelerate Diagnostics, USA
Krzysztof Zatwarnicki	Opole University of Technology, Poland
Zbigniew Zieliński	Military University of Technology, Poland
Liudong Zuo	California State University, USA
Piotr Zwierzykowski	Poznań University of Technology, Poland

Referees

Rajiv Bagai
Robert Bestak
Tomasz Bilski
Grzegorz Bocewicz
Leoš Bohac
Maria Carla Calzarossa
Lelio Campanile
Andrzej Chydziński
Dariusz Czerwiński
Adam Czubak
Peppino Fazio
Jean-Michel Fourneau
Natalia Gaviria
Agustín J. González
Anna Grocholewska-Czuryło
Daniel Grzonka
Artur Hłobaż
Mauro Iacono
Jacek Izydorczyk
Sergej Jakovlev
Anna Kamińska-Chuchmała
Ivan Kotuliak
Zbigniew Kotulski
Henryk Krawczyk
Udo R. Krieger
Rakesh Kumar
Andrzej Kwiecień

Piotr Lech
Wolfgang Mahnke
Marcin Markowski
Przemysław Mazurek
Jarosław Miszczak
Jolanta Mizera-Pietraszko
Evsey Morozov
Sasa Mrdovic
Gianfranco Nencioni
Remigiusz Olejnik
Michele Pagano
Oksana Pomorova
Sławomir Przyłucki
Tomasz Rak
Stefan Rass
Stefano Rovetta
Alexander Schill
Pavel Smolka
Ireneusz Smołka
Tomas Sochor
Jacek Stój
Kerry-Lynn Thomson
Ewaryst Tkacz
Homero Toral Cruz
Kurt Tutschku
Przemysław Włodarski
Grzegorz Zaręba

Sponsoring Institutions

Organizer

Faculty of Automatic Control, Electronics and Computer Science, Silesian University of Technology, Poland

Co-organizers

Faculty of Electronics, Telecommunications and Informatics, Gdańsk University of Technology, Poland

Committee on Informatics of the Polish Academy of Sciences, the Section of Computer Networks and Distributed Systems, Poland

Technical Partner

Technical Co-sponsor

IEEE, Poland

Conference Partner

iNEER, USA

Contents

Computer Networks

A Revised Gomory-Hu Algorithm Taking Account of Physical Unavailability of Network Channels

Winfried Auzinger[1]([✉]) [ID], Kvitoslava Obelovska[2] [ID],
and Roksolyana Stolyarchuk[3] [ID]

[1] Institute of Analysis and Scientific Computing, TU Wien, Vienna, Austria
winfried.auzinger@tuwien.ac.at
[2] Institute of Computer Science and Information Technologies,
Lviv Polytechnic National University, Lviv, Ukraine
obelyovska@gmail.com
[3] Institute of Applied Mathematics and Fundamental Sciences,
Lviv Polytechnic National University, Lviv, Ukraine
sroksolyana@yahoo.com
http://www.asc.tuwien.ac.at/~winfried/

Abstract. The classical Gomory-Hu algorithm aims for finding, for given input flows, a network topology for data transmission and bandwidth of its channels which are optimized subject to minimal bandwidth criteria. In practice, it may occur that some channels between nodes of the network are not active. Ignoring such channels using the topology obtained be the Gomory-Hu algorithm will not lead to an optimal flowrate.

In this paper the focus is on a modified algorithm taking into account deficient channels. While the classical algorithm generates a sequence of ring subnets, in our modified version the use of deficient channels is checked at intermediate stages in each cycle of the algorithm. When forming ring subnets, the availability of new channels to be introduced into the ring subnet is checked and in the case of unavailability another ring closest to the optimal one is selected. The network optimized by this modified algorithm guarantees the transmission of the maximum input stream.

Keywords: Network topology · Channel capacity · Gomory-Hu algorithm

1 Preamble

From a historical point of view, networking, data transmission and distributed processing developed as a result of scientific and technological progress. Modern networks connect a huge amount of computers and other devices via communication channels. In view of their growing complexity the problem of optimizing their performance has become more and more important.

© Springer Nature Switzerland AG 2020
P. Gaj et al. (Eds.): CN 2020, CCIS 1231, pp. 3–13, 2020.
https://doi.org/10.1007/978-3-030-50719-0_1

In mathematical terms, a network corresponds to a (weighted) graph, directional or non-directional. Network servers, routers, etc. are vertices, and communication lines (channels) are the edges of such a graph.

2   Introduction: Problems of Network Performance Optimization

A computer network is a complex combination of data terminal equipments, different data communication equipments such as routers, physical environment, channels, application processes, data flows, communication and routing protocols, etc. Therefore, optimization of network performance is a complicated task requiring from developers to assess the expected performance.

For studying such problems it is advisable not to create a real physical network[1] but to use a mathematical model, on the basis of which it will be possible to judge the efficiency of future networks and to decide whether to realize them or how to improve their topology.

A mathematical description of a network is typically based on a graph-theoretical model, where the set of vertices represents the nodes of the network, and the set of edges represents the channels connecting them. Network topologies are traditionally described as undirected graphs without loops or multiple edges. In Fig. 1 we visualize typical topologies.

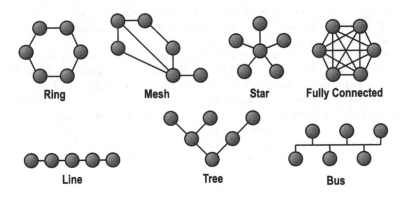

Fig. 1. Typical network topologies

---

[1] Even for smaller networks, physical modeling requires large effort, time and considerable material costs. Thus, the possibilities of physical modeling are rather limited. It only allows to investigate special settings where a small number of combinations of the relevant system parameters are taken into account.

The major criteria affecting efficiency are

- performance,
- reliability,
- security.

The performance of a network is affected by the number of users, transmission media, hardware and software limitations. It is measured in terms of

- transit time,
- response time,
- throughput,
- delay.

Finding an optimal topology for a designed network taking the above criteria into account, requires a multi-criteria analysis which is not easy to formalize. This is due to the fact that these criteria have different (often contradictory) effects on the analyzed object. For example, increasing reliability leads to redundant components, connections require better equipment, etc.

Many works were devoted to the design of different types of network topologies, examples of which can e.g. be found in [4, 6], where methods and algorithms for optimization are systematized and analyzed. The principles underlying the Gomory-Hu topological design are described in [2, 3, 6]. Gomory and Hu proposed an algorithm providing a synthesis of network topology and choice of channel capacities. The network designed according to this algorithm enables transmission of a maximum given input flow with a minimum required total capacity over the channels. In [7] a simulation model based on the Gomory-Hu algorithm (besides some other algorithms) was implemented. In [8], devoted to the embedding of virtual topologies in network clouds, the algorithmic steps proposed start with building the Gomory-Hu tree.

Optimization via the classical algorithm results in channel capacities which do not take into account particular transmission technologies in the different channels. But the capacity of the separate channels should be selected in accordance with the requirements of their transfer technology. An appropriately modified algorithm taking into account the requirements of the Dense Wavelength Division Multiplexing (DWDM) technology is presented in [1].

The aim of the present paper is to design a network topology with minimum excess capacity by modifying the Gomory-Hu algorithm for cases when a physical realization for some of the channels is not available or needs to be avoided.[2]

3   The Classical Gomory-Hu Algorithm Aiming for Optimizing the Network Topology and Selecting the Bandwidth of Its Channels

The input data is a set of nodes, together with the requirement of exchange of information and the intensity of flows that need to be provided between them. The algorithm assumes that this data are represented by a non-oriented weighted graph, The weights of the edges (=channels) correspond to the flows to be transmitted. The result of the optimization procedure is a weighted graph representing the topology of the network after optimization. The weights of channels represent the channel capacities.

Applying the classical Gomory-Hu algorithm one can find a network topology and the capacity of its channels for which transmission with maximum flow is ensured, and at the same time the weights of all edges (the required capacity of the communication channels) will be minimal. Let us discussed the application of the Gomory-Hu algorithm for network topology optimization and choosing capacity of its channels. The algorithm assumes that, according to certain rules, the input graph is divided into a set of graphs, of which represents some subnet. All subnets, with the exception of the last one, which can be a segment connecting two nodes, are ring subnets with the same weight value of each edge of the uniform ring. By superposition of all the resulting subnets one obtains an optimized network that will feature minimum total capacities of the channels (edges), while providing transmission of maximum flow. Found in the process of optimization the weights of the edges are equal to the required bandwidth of the channels. For better understanding the Gomory-Hu algorithm can be divided into two major stages:

(i) decomposition,
(ii) superposition.

---

[2] Nowadays there is a high risk of intentional damage caused by truncating communication lines. This problem is particularly relevant for backbone lines providers. For example, in 2017, there was the massive attack on Ukrtelecom's backbone lines serving eastern Ukraine. In two places a long-distance line was cut and cables were damaged [9].

According to the new General Data Protection Regulation (GDPR) [10] adopted in EU in 2018 it is prescribed to implement appropriate technical and organizational measures to ensure a level of security appropriate to the risk. In case of a physical or technical incident one should be able to recover all personal data in a timely manner. From our point of view, the best solution to provide these requirements when developing a backbone network topology is to exclude dangerous lines in the network, for example from the military area. Our modified algorithm will help to do this, although at the cost of slightly exceeding bandwidth.

(i) Decomposition can be described in following way:
1. Specification of a weighted non-directed graph G_{in}, in which vertices of graph represent network nodes, and edges represent flows where the weights of the edges represent the required intensities of flows.
2. Construction of the graph $A := G_{in}$.
3. Decomposition of the graph A into
    - a ring graph SN_i which includes all vertices of the graph A, which have edges and assigning to each edge of the ring the weight $W_{min}/2$; where i is the cycle number and W_{min} is the minimal weight of the edges of the graph A;
    - a graph B which is obtained by subtracting the value W_{min} from the weight of each edge of graph A whose weight is positive.
4. If the number of edges in the graph B is larger than one, we accept $A = B$ and go back to step 3; otherwise the decomposition is completed.
(ii) Superposition means constructing the output graph by integration of all graphs to which the input graph was decomposed in step 1. Let us illustrate the work of the algorithm on the example of a network topology optimization, network includes 7 nodes, among which should be transmitted streams described below in Example 1.

Example 1. The input data are the intensities of flows a_{ij} between the i-th and j-th nodes:

$$a_{12} \Rightarrow 100\,\text{Tb/s},$$
$$a_{14} \Rightarrow 50\,\text{Tb/s},$$
$$a_{15} \Rightarrow 20\,\text{Tb/s},$$
$$a_{23} \Rightarrow 6\,\text{Tb/s},$$
$$a_{25} \Rightarrow 10\,\text{Tb/s},$$
$$a_{46} \Rightarrow 4\,\text{Tb/s},$$
$$a_{67} \Rightarrow 10\,\text{Tb/s}.$$

Figure 2 shows the process and result of optimization based on the classical Gomory-Hu algorithm.

Let us discuss these results. The verification is performed for maximum flow, in our case it is $a_{12} = 100$ Tb/s. In the resulting network, this flow can be transmitted simultaneously in the following way:

$$a_{12} \Rightarrow 75\,\text{Tb/s},$$
$$a_{1765432} \Rightarrow 3\,\text{Tb/s},$$
$$a_{176542} \Rightarrow 2\,\text{Tb/s},$$
$$a_{1542} \Rightarrow 5\,\text{Tb/s},$$
$$a_{142} \Rightarrow 15\,\text{Tb/s}.$$

Here, $a_{1765432}$ denotes the path from node 1 to node 2 across nodes 7, 6, 5, 4 and 3, etc. The sum of these flows is 100 Tb/s. Thus, since the maximum input

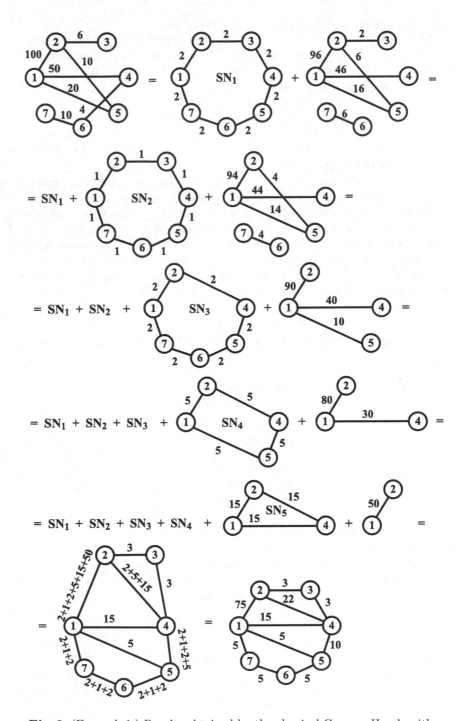

Fig. 2. (Example 1:) Results obtained by the classical Gomory-Hu algorithm.

stream is 100 Tb/s the network is able to transmit the maximum flow without any excess resources. i.e., all network resources are busy for transmission of maximum flow. The total capacity of all channels is 148 Tb/s.

Consider the case where for some reason it is not possible to provide some channel, or channels connecting the nodes of the outer ring. In this case it will not be possible to transmit the maximum input stream. The reason for this is that the algorithm provides an optimal solution which eliminates any redundancy. Therefore, rejecting any channel or reducing its bandwidth will not permit to achieve the desired goal.

4   A Modified Gomory-Hu Algorithm for Optimizing the Network Topology and Selecting the Bandwidth of Its Channels

Now we propose a modification of the classical algorithm in order to select the network topology and bandwidths of its channels according to the criteria of minimum bandwidth, in case of possible restrictions on the existence of certain channels. Additional input is a 'black list' of unavailable channels between the nodes of the outer ring. In the classical algorithm, ring subnets are iteratively constructed in course of the decomposition. In the modified algorithm, in each cycle it is checked whether a channel from the black list of unavailable channels is included in the ring subnet. If such a channel is not contained in the black list, the algorithm proceeds in the same way as the classical one.

Otherwise, we propose to use at this step of the algorithm the topology of the previous ring. Obviously, there will be some redundancy, but it will offer a solution that can be implemented. The modified algorithm can be described as follows.

1. Specification of a weighted non-directed graph G_{in}, in which vertices of graph is network nodes and edges are flows, weights of the edges represent the required intensities of flows.
2. Construction of graph $A := G_{in}$.
3. Find the minimal weight of the edges of the graph $A - W_{min}$.
4. Decomposition of graph A into:
   - a ring graph SN1 which includes all the nodes of graph A, and assigning to each edge of the ring the weight $W_{min}/2$;
   - a graph B, which is obtained by subtracting the value W_{min} from all edges of graph A whose weight is positive.
5. If the number of edges in graph B is equal one, then go to step 11, else we accept $A = B$.
6. Construction a ring graph SN_k which includes all the nodes of graph A, where k is the cycle number, which started from 2.
7. Comparing ring graphs $SN_k = SN_{k-1}$ to detect new edge. If the new edge does not appear go to 8. Otherwise check if the new edge is forbidden according to the list of forbidden channels. In this case, set $SN_k = SN_{k-1}$.

8. Find the minimal weight of the edges of the graph $A - W_{min}$.
9. Decomposition of graph A into:
   - a ring graph SN_k and assigning to each edge of this ring the weight $W_{min}/2$;
   - a graph B which is obtained by subtracting the value W_{min} from all edges of graph A whose weight is positive.
10. If the number of edges in graph B is larger than one, then we accept $A = B$ and go back to step 6.
11. Integration of all graphs of SN_k and graph B.

Example 2 (based on Example 1:). An illustration of the proposed approach is shown in the example of the implementation of the modified algorithm for the same input data as before for the classical version. Figure 3 shows the process and result of optimization based on our modified algorithm.

The verification is performed for maximum flow, in our case it is $a_{12} = 100$ Tb/s. In the resulting network, this flow can be transmitted simultaneously in the following way:

$$a_{12} \Rightarrow 75 \text{ Tb/s},$$
$$a_{1765432} \Rightarrow 3 \text{ Tb/s},$$
$$a_{176542} \Rightarrow 7 \text{ Tb/s},$$
$$a_{142} \Rightarrow 15 \text{ Tb/s}.$$

The sum of these flows is 100 Tb/s. Thus, this network also is able to transmit the maximum flow. The connection between nodes 1 and 5 is not involved, which cannot be physically implemented due to input conditions. The price for this is increasing the total capacity of all channels. In this example it equals 158 Tb/s. When optimizing according to the classical Gomory-Hu algorithm the total capacity of all channels was 148 Tb/s, but the classical algorithm does not take into account the constraints caused by unavailability of some channels.

For automated verification of results, the Gomory-Hu algorithm was used to determine the maximum flow in the network. The following are fragments from an implementation of the program for finding the maximum streams for the examples discussed above. The results for Examples 1 and 2 are shown in Fig. 4 and 5, respectively.

In the input data for the classical and modified algorithms, the maximum flow value to be transmitted between nodes 1 and 2 is 100 Tb/s. As can be seen from Figs. 4 and 5, the optimized topologies provide exactly this value.

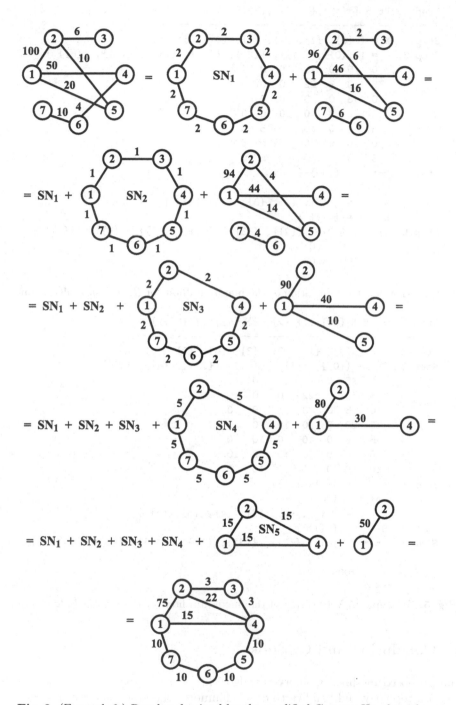

Fig. 3. (Example 2:) Results obtained by the modified Gomory-Hu algorithm.

```
Step 1: Vt = [ {{0},{1},{2},{3},{4},{5},{6}} ]
------------------------------------------------
Step 2: X = [ {0}, {1}, {2}, {3}, {4}, {5}, {6} ]
Step 3: G = [ {{0}}, {{1}}, {{2}}, {{3}}, {{4}}, {{5}}, {{6}} ]
             0  75  0  15   5  0  5
            75   0  3  22   0  0  0
             0   3  0   3   0  0  0
            15  22  3   0  10  0  0
             5   0  0  10   0  5  0
             0   0  0   0   5  0  5
             5   0  0   0   0  5  0
Step 4: s-t   = {{0}}-{{1}}
        max_f = 100
        A     = [ {0}, {2}, {3}, {4}, {5}, {6} ]
        B     = [ {1} ]
Step 5: Vt = [ {{0}}, {{2}}, {{3}}, {{4}}, {{5}}, {{6}},  {{1}} ]
               0  100
             100   0
```

Fig. 4. (Example 1:) Verification of the maximum flow for the classical algorithm.

```
Step 1: Vt = [ {{0},{1},{2},{3},{4},{5},{6}} ]
------------------------------------------------
Step 2: X = [ {0}, {1}, {2}, {3}, {4}, {5}, {6} ]
Step 3: G = [ {{0}}, {{1}}, {{2}}, {{3}}, {{4}}, {{5}}, {{6}} ]
             0  75  0  15   0   0  10
            75   0  3  22   0   0   0
             0   3  0   3   0   0   0
            15  22  3   0  10   0   0
             5   0  0  10   0  10   0
             0   0  0   0  10   0  10
            10   0  0   0   0  10   0
Step 4: s-t   = {{0}}-{{1}}
        max_f = 100
        A     = [ {0}, {2}, {3}, {4}, {5}, {6} ]
        B     = [ {1} ]
Step 5: Vt = [ {{0}}, {{2}}, {{3}}, {{4}}, {{5}}, {{6}},  {{1}} ]
               0  100
             100   0
```

Fig. 5. (Example 2:) Verification of the maximum flow for the modified algorithm.

5   Conclusion and Outlook

The proposed modified Gomory-Hu algorithm offers a solution for finding the network topology and bandwidth of its channels, when for some reasons certain channels are not available. A network optimized in this way still guarantees the transmission of the maximum input stream.

Further work on this topic (see [5] and Acknowledgement below) will include a study of complexity issues, also based on a systematic implementation, preferably using the graph-theoretical features available in the computer algebra system Maple.

Acknowledgement. (i) This work was realized within the framework of the Jean Monnet 2019 program under Erasmus+'611692-EPP-1-2019-1-UAEPPJMO-MODULE Data Protection in EU'.

It is also to be considered as a first step in a cooperation project under preparation, following up the joint Ukraine-Austria R & D project 'Traffic and telecommunication networks modelling', project No. UA 10/2017/0118U001750, M-130/2018 (cf [1]).

(ii) The authors thank Nadyja Popovuch, student at Lviv Polytechnic National University, for providing figures.

References

1. Auzinger, W., Obelovska, K., Stolyarchuk, R.: A modified Gomory-Hu algorithm with DWDM-oriented technology. In: Lirkov, I., Margenov, S. (eds.) LSSC 2019. LNCS, vol. 11958, pp. 547–554. Springer, Cham (2020). https://doi.org/10.1007/978-3-030-41032-2_63
2. Ford, L.R., Fulkerson, D.R.: Flows in Networks. Princeton University Press, Princeton (1962)
3. Gomory, R.E., Hu, T.C.: Multi-terminal network flows. J. Soc. Ind. Appl. Math. 9(4), 551–570 (1961)
4. Kasprzak, A.: Exact and approximate algorithms for topological design of Wide Area Networks with non-simultaneous single commodity flows. In: Sloot, P.M.A., Abramson, D., Bogdanov, A.V., Gorbachev, Y.E., Dongarra, J.J., Zomaya, A.Y. (eds.) ICCS 2003. LNCS, vol. 2660, pp. 799–808. Springer, Heidelberg (2003). https://doi.org/10.1007/3-540-44864-0_82
5. King, V.: Fully dynamic connectivity. In: Kao, M.Y. (ed.) Encyclopedia of Algorithms. Springer, Boston (2015)
6. Larsson, C.: Design of Modern Communication Networks. Methods and Applications. Elsevier Ltd., Amsterdam (2014)
7. Tapolcai, J.: Routing algorithms in survivable telecommunication networks. Ph.D., Thesis Summary, Budapest, Hungary (2004)
8. Xin, Y., Baldine, I., Mandal, A., Heermann, C., Chase, J., Yumerefendi, A.: Embedding virtual topologies in networked clouds. In: CFI 2011 Proceedings of the 6th International Conference on Future Internet Technologies, Seoul, Korea, pp. 26–29. ACM New York (2011)
9. https://www.rbc.ua/ukr/news/ukrtelekome-zayavili-massovoy-atake-magistr alnye-1494495888.html
10. Regulation (EU) 2016/679 of the European Parliament and of the Council of 27 April 2016 on the protection of natural persons with regard to the processing of personal data and on the free movement of such data, repealing Directive 95/46/EC (General Data Protection Regulation). https://eur-lex.europa.eu/legal-content/EN/TXT/PDF/?uri=CELEX:32016R0679

Analysis of the Block Segmentation Method of the Licklider Transmission Protocol

Ricardo Lent[✉]

University of Houston, Houston, TX, USA
rlent@uh.edu

Abstract. Space communications are continuously challenged by extreme conditions that include large propagation delays, intermittent connectivity, and random losses. To combat these problems, the Licklider Transmission Protocol (LTP) splits data blocks into small segments that are radiated independently and retransmitted as needed, through a process that can be paused during long link disruptions. Given the extreme delays involved, the end performance of this protocol is driven by the number of transmission rounds needed to successfully deliver each block. LTP links are defined as overlays with one or more physical channels in the underlay, therefore with sections that may be on different administrative domains and experiencing different conditions. The question of how to select the length of the segments has received negligible attention and the use of improper values can easily lead to suboptimal performance. The segmentation process used by LTP is examined in this paper to determine the role that segmentation parameters and the conditions of the underlay have on the block delivery times. This goal is achieved through the definition of a basic model of LTP's transmission process that allows deriving the optimal segmentation parameter. Simulation results provide additional evidence of LTP's performance contrasting the results of the optimal segment length with fixed-length segments. The results provide a theoretical performance reference for practical parameter optimization methods.

Keywords: Delay tolerant networks · Deep space communications · Satellites · Licklider Transmission Protocol · Protocol optimization

1 Introduction

Space communications networks are made distinctive by the use of long-distance links, which both entail extreme one-way propagation light times (OWLT) and long down periods. For example, the OWLT for Earth-Moon communications is around 1.2 s whereas for the one for Earth-Mars is in the range of 4–24 min. Orbital mechanics also bring occultation (i.e., celestial bodies blocking communications), which can prevent any communication from a few minutes to

© Springer Nature Switzerland AG 2020
P. Gaj et al. (Eds.): CN 2020, CCIS 1231, pp. 14–26, 2020.
https://doi.org/10.1007/978-3-030-50719-0_2

several hours. Along with the Bundle Protocol (BP), the Licklider Transmission Protocol was developed specifically to address those issues. LTP defines overlay links with support for both best-effort and reliable data delivery. Because of the use of network abstraction, LTP links are built on top of another protocol as desired without constraint to any specific layer. Therefore, the underlay could be selected to be a link-layer protocol or a higher-layer protocol, such as TCP or UDP.

In the current state-of-practice, space radio channels and the link-layer are regularly tuned to improve their performance, e.g., by redesigning link budgets or through the use of error correction codes and buffer management techniques. However, the problem of how to optimize LTP-specific parameters has received relatively low attention in the past. The ION-DTN [1] implementation of LTP provides the means for the manual configuration of different protocol parameters, such as the maximum block size, number of import and export sessions (which controls the transmission concurrency), and the maximum segment size. However, limited automation for finding the optimal values for these parameters is available, requiring many times the use of human expertise.

In this paper, the segmentation method used by LTP is analyzed, which consists of dividing a data block into small chunks (i.e., segments) for transmission. In all current implementations, including ION-DTN, the maximum segment length is manually set usually matching the underlay's maximum transmission unit (MTU). The hypothesis is that the current approach for setting the segment length may not yield optimal performance for all situations as large segments tend to be more error-prone than smaller segments. Lost segments need to be retransmitted, which extends the delivery time for the block. However, the use of small segments may not be desirable either as they add header overhead both from LTP and the underlay. Therefore, the tradeoffs in the selection of the segment length require further examination.

The contributions of this paper are three-fold: (1) it analyzes LTP's block segmentation process, yielding a model that describes the role of the conditions of the underlay network and the selected protocol parameters to the overlay link efficiency, (2) it finds the theoretical ideal segment length that maximizes link efficiency and that leads to lower response times and higher throughout than achievable through the common practice of using fixed segment sizes, and (3) it provides the ideal LTP performance that serve as a reference for future works.

2   Related Works

A common criterium for determining the size of the data units in a network is the possibility of fragmentation. As stated in RFC 2488, which defines performance enhancements for satellite channels using standard network mechanisms, it is recommended "the use of the largest packet lengths that prevent fragmentation", which can be determined by the Path MTU Discovery (RFC 1191) mechanism. In addition, it recommends the use of forward error correction (FEC) to reduce the possibility of triggering congestion control actions when TCP is used.

Several works have suggested the practical link between the selected packet sizes and performance, in particular, considering challenged networks. For example, Basagni et al. studied the impact of the packet size selection in an underwater wireless sensor network [2] showing that the performance of the carrier sense multiple access (CSMA) and the distance-aware collision avoidance protocol (DACAP) is impacted by the packet size selection even with BER values as low as 10^{-6}.

Space networks are challenged networks that are especially exposed to the impact of packet losses due to the long propagation delay of the links. This observation has been well documented by numerous studies [3–6] that have focused on different aspects of LTP, such as flow control [7], the aggregation of bundles in blocks [8,9] and the impact of the selection of the convergence layer [10,11]. Recent works have focused on enhancing LTP for performance gains [12–14], for example through the use of a Reed-Solomon code [15] to reduce the segment loss probability.

An experimental work, conducted by Bezirgiannidis and Tsaoussidis [16], measured the effect that packet sizes have on LTP performance. Several works have looked into the properties of the radio channel and the formulation of the packet optimization problem (e.g., see [17,18]). Close work was carried out by Lu et al. who analyzed the approximated impact of packet sizes to LTP's performance and formulated a heuristic to find the optimal length [19]. Because of the use of a performance model, a drawback of their approach is that it requires knowledge of the channel state which may not be available given the overlay nature of LTP links and the possible lack of cross-layer information. In recent work, a cognitive networking approach to the dynamic selection of the optimal segment length that does not require precise knowledge of the underlay was also proposed [20].

3   Synopsis of the Licklider Transmission Protocol

The Licklider Transmission Protocol (LTP) [21,22] was introduced as a convergence layer protocol for the Bundle Protocol (BP) to support bundle transmissions over one or multiple links (as an overlay) that are expected to be disrupted for extended times. LTP receives and transmits *service data units* from and to BP (i.e., data bundles). As data arrives from BP, the sending LTP engine accumulates the data in a buffer to create a data block. Each data block is segmented and sent independently of each other to the receiving LTP engine. To this end, LTP uses lower layer protocols. Typically, CCSDS/AOS (Consultative Committee for Space Data Systems/Advanced Orbit System) for single-hop radio links and internet protocols (UDP, TCP, STCP) for overlays. The size of the data blocks is determined by two parameters, which limit the amount of memory reserved for each block and the filling time (typically, 1 s) respectively. Therefore, the actual size of any particular block is not necessarily fix and may include a whole bundle, part of a bundle, or even multiple bundles.

LTP supports both best-effort and reliable transmissions within the same block. The portion of the block devoted to best-effort is labeled the *green-part*, whereas the *red-part* is reserved for the reliable part. This division is possible because each block is segmented and each block transmission consists in sending a sequence of data segments (DS). This process is called a *session*. The size of a segment is commonly set to match the value of the maximum transmission unit (MTU) of lower-layer protocol(s), but the benefits of selecting smaller segments based on the context are suggested in this paper. The protocol does not put any particular limit on the size of the segments, so that segments may be smaller than the MTU or even larger (therefore, spanning multiple data-link frames).

Each session requires at least one transmission round. During that round all of the block's DS are sent. The last segment is labeled the *checkpoint segment* (CP) and carries the flag *end-of-red-part* (EORP). This flag tells the receiver to respond with a report segment (RS). The RS provides negative acknowledgment to the sender so that it can retransmit lost segments. Additional rounds can proceed until all of the segments are either successfully received or the maximum number of rounds is reached. In the latter case, the block is discarded. On reception of the final confirmation (RS) from the receiver, the sender transmits an acknowledge (RA) to close the block transmission. LTP blocks may be transmitted one at a time, i.e., transmission of a block starts only after the previous completely finishes, or in multiple concurrent sessions. The latter mode offers better performance but may be limited by the memory available to LTP at both ends. From the standpoint of the sender, the session finishes as soon as the RA is transmitted. Figure 1 illustrates the process taking place as an overlay link.

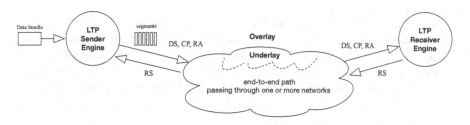

Fig. 1. An LTP link defines a network abstraction (overlay) that is not restricted to a single physical link, but that runs on top of an arbitrary underlay such as UDP/IP.

4   Analysis

To analyze the theoretical impact of the segmentation process, the study first looks at the nominal (i.e., error-free) overlay link efficiency and capacity and then extends the results to the lossy case. Capacity refers to the upper bound of the rate at which bundles can be reliably received.

The study considers the general case of application data being reliably sent over an overlay link by BP over LTP. The overlay is built over an underlay of Z channels, that is, the underlay path consists of Z physical links. In the discussion that follows, a *frame* is the name of the underlay protocol data unit.

Despite the usual association of that term to the link layer, no specific restrictions are assumed about the underlying protocol, so a *frame* could be a UDP datagram for instance or a CCSDS frame.

4.1 Nominal Bundle Capacity

Each bundle consists of at least two structures. One structure carries the payload, whereas the second, and additional structures if any, carries control information. Let us use B to represent the total bundle length and h_B the total length used by the control structures within the bundle. The mapping of bundles to LTP blocks depends on the amount of application data available to be sent, implementation specifics, and protocol configuration. It may result in blocks carrying single bundles, a fraction of a bundle, or multiple bundles. To model these alternatives, let use parameter a to indicate the *bundle-block aggregation factor*: $a = B/b$, where b is the block size. If $a = 1$, then one block carries exactly one bundle; if $a > 1$, multiple blocks ($\lceil a \rceil$ blocks) carry one bundle; and, if $a < 1$, one block aggregates different bundles. Furthermore, let m and h represent the segment payload length and header length respectively. To simplify the notation, let us assume that h includes the lengths of both the segment and link-layer frame so that the total segment length at the physical layer is $L = m + h$. Protocol extensions, such as security mechanisms extensions [23], involve the addition of extra control information and so, lead to a larger value for h. Each block requires the transmission of $n = b/m$ segments or frames, since it is assumed that each segment travels in one frame.

The nominal BP/LTP efficiency is the ratio of the maximum BP payload length to the total transmission length including overhead and data. To calculate the protocol efficiency, it can be noted that the total overhead of a bundle transmission over LTP is $h_B + anh$ given that each bundle introduces control overhead (h_B) and involves the transmission of a blocks of n segments. Each segment within each block adds a separate protocol overhead h. The nominal BP/LTP efficiency $\mathcal{E}_{nom}$ is, therefore:

$$\mathcal{E}_{nom} = \frac{B - h_B}{B + ahn} = \frac{1 - h_B/B}{1 + h/m} \tag{1}$$

The values of both h_B and h are comparatively fixed, and the value of B depends on the application requirements. The numerator in (1), $1 - h_B/B$, represents the bundle efficiency and varies between 0 and 1. Increasing the bundle overhead h_B decreases bundle efficiency linearly. Values of B of at least 10 times larger than h_B allow achieving a bundle efficiency of 0.9 or higher. The denominator in (1), $1 + h/m$, is the frame overhead factor. The expression indicates that larger packet payload m helps to reduce the frame overhead and also protocol efficiency. It can also be observed that realistically, m is the only controllable variable that affects the nominal BP/LTP protocol efficiency. Another observation is that the block size b and the aggregation factor a do not impact nominal protocol efficiency.

The nominal link overlay capacity is the product of its efficiency and the end-to-end throughput r. With the assumption of negligible buffering delays in this work, the processing time of frames at the different sections of the overlay are only determined by the links' rates. Therefore, the overlay link's throughput is given by the slowest link of the underlay path. Furthermore, let us denote with v the overlay path availability that results due to link disruptions (e.g., as caused by occultation or link scheduling actions in the space communications environment). The nominal capacity θ_{nom}^+ is given by: $\theta_{nom}^+ = \mathcal{E}_{nom}vr$.

4.2   Reliable Bundle Efficiency and Overlay Link Capacity

Consider now the reliable transmission of bundles over a lossy overlay link that randomly drops frames with probability p. Because the overlay may consist of multiple sections, a number of these sections may implement forward error codes, such as Reed-Solomon coding or Turbo coding, that can help to mitigate frame drops. Also, additional packet encapsulation and multiplexing may occur prior to the physical layer transmission at each section. Parameter p then models the resulting end-to-end packet drop probability after considering any error mitigation technique used in all sections of the underlay. In detail, let b_i denote the bit error rate (BER) of the i-th channel of the underlay. For a frame of size L with independent bit errors, the packet error rate p_i for section i, $i = 1, \ldots Z$, where Z is the underlay path length, is given by the expression:

$$p_i = 1 - (1 - b_i)^L = 1 - e^{L.log(1-b_i)} \tag{2}$$

Because frames could be dropped at any section of the underlay, parameter p is then given by the expression:

$$p = 1 - \prod_{i=1}^{Z}(1 - p_i) \tag{3}$$

It is relevant to emphasize that in practice frames may be also dropped due to congestion. The model in this paper only considers channel losses and a negligible probability of buffer overflow. Assuming that each segment travels on a separate frame, the packet loss probability p will affect the transmission capacity of the overlay θ_s^+ as follows:

$$\theta_s^+ = (1 - p)vr/L \tag{4}$$

given that the maximum segment throughput of r/L is also limited by the overlay link availability v and only the fraction $(1-p)$ of the segments will not be rejected on average due to errors. Because each LTP block consists of n segments, the maximum reliable block transmission capacity (θ_b^+) becomes:

$$\theta_b^+ = \theta_s^+/n. \tag{5}$$

Despite LTP implements reliable block transmission, a block may still be dropped after a certain number of unsuccessful delivery attempts. It is generally beneficial to decide first the value of the target block loss rate (α) and then calculate that maximum number of rounds that can achieve such a level [24]. It is considered that α is known as is given as an upper bound by design and that $\alpha \ll 1$, which is typical and allows ignoring the impact of the maximum number of rounds to the average value (κ).

If the bundle fits within one LTP block, then the probability of losing the bundle is just α. However, if the bundle involves several block transmissions, then at least one block loss causes the entire bundle rejection. A bundle is not lost with probability $(1 - \alpha)^A$, where $A = max\{1, a\}$, where a is the bundle-block aggregation factor previously defined. The maximum reliable bundle capacity (θ_B^+) is then:

$$\theta_B^+ = (1 - \alpha)^A \theta_b^+ / a. \tag{6}$$

Observing that each bundle carries $(B - h_B)$ of useful data, and using (4), (5) and (6), it can be determined the reliable capacity of BP/LTP as:

$$\theta^+ = (B - h_B)\theta_B^+ = \mathcal{E}rv/a \tag{7}$$

where

$$\mathcal{E} = \frac{1 - h_B/B}{1 + h/m}(1 - \alpha)^A(1 - p) = \mathcal{E}_{nom}(1 - \alpha)^A(1 - p). \tag{8}$$

The reliable efficiency $\mathcal{E}$ is equivalent to the nominal efficiency subject to not losing a segment nor a bundle. Unlike the nominal efficiency case, parameter m not only impacts the segment overhead factor in the reliable efficiency, but also the segment loss probability p (3). Expression (8) also suggests that the transmission of a bundle using multiple blocks decreases the reliable efficiency.

Figure 2 (a) compares the nominal BP/LTP efficiency with the reliable BP/LTP efficiency for a range of values of the segment payload length m and under different BER conditions. The model parameters include 100 kB bundles with 100 B control overhead, an LTP header length of 12 B, one block per bundle ($a = 1$), and a block loss rate of 10^{-6}. At low packet loss rates, the nominal and reliable overlay link efficiencies are very similar. The situation changes at high packet loss rates in particular with large segments.

4.3   Optimal Payload Length

As previously indicated (7), only the efficiency factor that appears in the overlay link capacity depends on the segment payload length m. The overlay link throughput (r), overlay availability (v), and bundle-block aggregation factor (a) are insensitive to the choice of the segment payload length. The optimal value of the segment payload length m^* yields the largest $\mathcal{E}$, which can be found by solving:

$$\underset{m}{\text{maximize}} \quad \mathcal{E}(m)$$

$$\text{subject to} \quad 0 < m \leq B, M$$

$$m \in \mathbb{Z}.$$

where M is the smallest payload-length that is supported along the path of the underlay network. The single-section network case (i.e., single-hop, $b = b_1$) leads to a simple expression that helps to illustrate the tradeoffs involved in the segment length selection. It can be found by temporarily ignoring the two constraints and finding the horizontal tangent with $\mathcal{E}' = 0$:

$$m^+ = \frac{h}{2}\left(1 + \sqrt{1 - \frac{4}{h\,log(1-b)}}\right).\tag{9}$$

The constraints can then be applied using: $m^* = max\{1, min\{\lfloor m^+ \rfloor, B, M\}\}$. Expression (9) reinforces the notion that larger segments are generally better suited for low BER with the opposite case otherwise.

Figure 2 (b) depicts the reliable efficiency achieved by the optimal payload under different BER values, compared to fixed payloads of 2, 20, 200, and 2000 bytes. The case corresponds to a bundle size of 10^5, single bundle carrying blocks, and bundle and frame overhead of 100 and 12 bytes respectively. The probability of block loss was fixed at 10^{-6}. The results illustrate the tradeoffs between header overhead and segment drops involved in the selection of the LTP segment length.

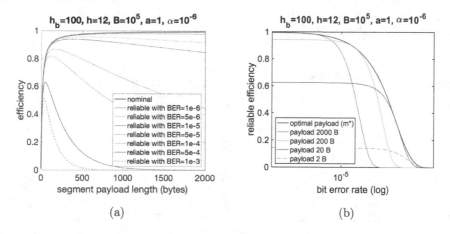

Fig. 2. (a) Nominal vs. reliable BP/LTP protocol efficiency, and (b) efficiency achieved with fixed segments of various lengths (i.e., LTP) and with the optimal segmentation.

5   Simulation Results

To verify the performance achievable by the optimal segmentation, a discrete event simulator of BP/LTP was used. The performance was measured in terms of the block delivery time, response time, and bundle throughput. The simulator keeps track of the timing of all the packet buffering and transmission events, in addition to the packet drops, of a packet flow that is being carried by a single

256-kbps wireless channel with given one-way light time (OWLT) and bit error rate (BER). The accuracy of the simulator has been verified previously [24,25].

The OWLT and BER channel values were used as experimental factors. It is assumed that the link was always available ($v = 1$). Statistics of the transmission of 1000 bundles of size 100 kB were collected to characterize LTP performance. Identical parameters to the theoretical model were assumed in the simulation: a bundle fits exactly one block ($a = 1$) and the block control overhead was fixed to 100 bytes. In addition, each segment carries a header of 12 bytes and RS/RA (report acknowledgement) are 20 and 7 bytes long respectively.

5.1 Optimal Segment Size

Figure 3 (a) depicts the average segment length as determined by expression 9. The independent parameter is the BER value of the single channel used in the simulations. As a reference, the results obtained with fixed segment lengths of 500, 1000, and 1400 B have been included. These values are within the typical range used in practice, for example, RFC 879 states that the default maximum segment size (MSS) for TCP is 536 B and the general recommendation for LTP in the ION-DTN implementation is the use of 1400 B. The evaluation included an intermediate segment length between these two extremes. The optimal segment length was constrained to the range 10–1400, which explains the constant optimal value for BER values $<10^{-6}$.

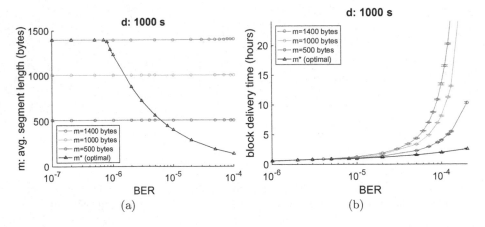

Fig. 3. (a) Optimal segment lengths and (b) block delivery times.

Figure 3 (b) depicts the average bundle delivery times obtained with the optimal segmentation and the fixed segment lengths. The performance advantage of adjusting the segment length become apparent for BER values higher than 10^{-5}. In the simulations, no error-correcting code was assumed. However, the coding gain of real channels is expected to simply shift the results along the horizontal axis.

5.2 Average Block Response Time and Throughput

Unlike the block delivery time (i.e., service time), the block response time is affected by the block sending rate and buffering. Higher sending rates increase the chances of buffer congestion extending response times. The effect is depicted in Fig. 4 (a) with a BER of 10^{-5} and for a range of values for the sending rate up to 2.3×10^{-3} bundle/s. It is worth noting that the values that are shown in the chart are not steady-state averages, but the average response time for 1,000 bundle transmissions at the selected rate. The benefits become particularly significant, measured in the range of hours for the 1,000 s link, as bundles are sent at higher rates. The optimal segmentation achieves up to 20–30% higher throughput than with the use of fixed segment lengths as depicted in Fig. 4 (b).

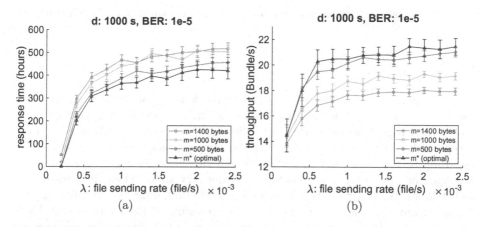

Fig. 4. Bundle response time (a) and throughput (b) over a single channel with a BER value of 10^{-5} and one-way propagation delay (d) of 1,000 s as a function of the bundle sending rate. The values were calculated for the first 10^3 bundles (not steady state).

5.3 Impact of the Propagation Delay

While the propagation delay does not affect the average number of rounds required to delivery a block by LTP, the time required for each round is a function of the one-way propagation delay of the channel. The relationship between block delivery time and propagation delay is linear as it can be observed in Fig. 5 for BER values of 10^{-5} and 10^{-4}. This evaluation covers a wide range of OWLT, from terrestrial to values associated to links to the edge of the solar system.

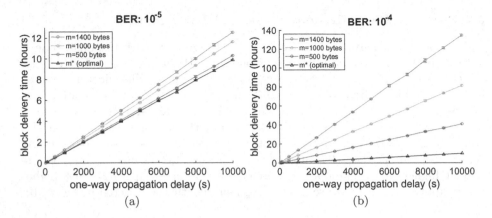

Fig. 5. Block delivery time vs. one-way propagation delay with channel BER values of (a) 10^{-5} and (b) 10^{-4}.

6   Conclusion

In this paper, the connection between bundle delivery performance and both the overlay link properties and the choice of the segmentation parameter was analyzed. A model of the overlay link efficiency was proposed and used to find the optimal segment length. Contrary to the default practice of setting the length according to the maximum transmission unit of the underlay, the results show that such practice may lead to suboptimal delivery times when the end-to-end packet loss ratio is high. Given that end-to-end communication conditions can change at any time, it can be inferred that an online method for defining LTP parameters should be employed rather than keeping this parameter constant as commonly done in practice today. Moreover, the overlay nature of LTP links may complicate the optimization as multiple physical-layer channels may be involved in the underlay negating the possible benefits of static techniques that try to map BER measurements to segment lengths. Nevertheless, it is expected that the results of this study will help to define a performance reference for future practical methods for LTP parameter optimization.

Acknowledgment. The author would like to thank Gilbert Clark at NASA Glenn Research Center for his useful comments on this research. This work was supported by an Early Career Faculty grant from NASA's Space Technology Research Grants Program.

References

1. The Interplanetary Overlay Network (ION) software distribution: ION-DTN. https://sourceforge.net/projects/ion-dtn. Accessed 01 Apr 2018
2. Basagni, S., Petrioli, C., Petroccia, R., Stojanovic, M.: Optimized packet size selection in underwater wireless sensor network communications. IEEE J. Ocean. Eng. **37**, 321–337 (2012)

3. Bisio, I., Marchese, M.: Analytical expression and performance evaluation of TCP packet loss probability over geostationary satellite. IEEE Commun. Lett. **8**(4), 232–234 (2004)
4. Caini, C., Cornice, P., Firrincieli, R., Livini, M., Lacamera, D.: Analysis of TCP and DTN retransmission algorithms in presence of channel disruptions. In: 2009 First International Conference on Advances in Satellite and Space Communications, pp. 174–179, July 2009
5. Wang, R., Burleigh, S.C., Parikh, P., Lin, C.J., Sun, B.: Licklider Transmission Protocol (LTP)-based DTN for cislunar communications. IEEE/ACM Trans. Netw. **19**(2), 359–368 (2011). http://dx.doi.org/10.1109/TNET.2010.2060733
6. Wang, D.: Performance of Licklider Transmission Protocol (LTP) in LEO-satellite communications with link disruptions. In: 2016 IEEE 15th International Conference on Cognitive Informatics Cognitive Computing (ICCI*CC), pp. 154–159, August 2016
7. Wang, R., Reshamwala, A., Zhang, Q., Zhang, Z., Guo, Q., Yang, M.: The effect of "window size" on throughput performance of DTN in lossy cislunar communications. In: 2012 IEEE International Conference on Communications (ICC), pp. 68–72, June 2012
8. Bezirgiannidis, N., Burleigh, S., Tsaoussidis, V.: Delivery time estimation for space bundles. IEEE Trans. Aerosp. Electron. Syst. **49**(3), 1897–1910 (2013)
9. Wei, Z., Wang, R., Zhang, Q., Hou, J.: Aggregation of DTN bundles for channel asymmetric space communications. In: 2012 IEEE International Conference on Communications (ICC), pp. 5205–5209, June 2012
10. Wang, R., Modi, B., Zhang, Q., Hou, J., Guo, Q., Yang, M.: Use of a hybrid of DTN convergence layer adapters (CLAs) in interplanetary Internet. In: ICC, pp. 3296–3300. IEEE (2012)
11. Wang, R., Wu, X., Wang, T., Liu, X., Zhou, L.: TCP convergence layer-based operation of DTN for long-delay cislunar communications. IEEE Syst. J. **4**(3), 385–395 (2010)
12. Farrell, S., Cahill, V.: Evaluating ltp-t: a DTN-friendly transport protocol. In: 2007 International Workshop on Satellite and Space Communications, pp. 178–181, September 2007
13. Alessi, N., Burleigh, S.C., Caini, C., de Cola, T.: LTP robustness enhancements to cope with high losses on space channels. In: 8th Advanced Satellite Multimedia Systems Conference and the 14th Signal Processing for Space Communications Workshop, ASMS/SPSC 2016, Palma de Mallorca, Spain, 5–7 September 2016, pp. 1–6 (2016)
14. Iannicca, D., Hylton, A., Ishac, J.: A performance evaluation of NACK-oriented protocols as the foundation of reliable Delay-Tolerant Networking convergence layers. NASA technical memorandum (2012)
15. Shi, L., et al.: Integration of reed-solomon codes to Licklider Transmission Protocol (LTP) for space DTN. IEEE Aerosp. Electron. Syst. Mag. **32**(4), 48–55 (2017)
16. Bezirgiannidis, N., Tsaoussidis, V.: Packet size and DTN transport service: evaluation on a DTN testbed. In: International Congress on Ultra Modern Telecommunications and Control Systems, pp. 1198–1205, October 2010
17. Carek, D.A.: Packet-based protocol efficiency for wireless communications. JACIC **2**, 238–251 (2005)
18. Korhonen, J., Wang, Y.: Effect of packet size on loss rate and delay in wireless links. In: IEEE Wireless Communications and Networking Conference, WCNC, vol. 3, pp. 1608–1613, April 2005

19. Lu, H., Jiang, F., Wu, J., Chen, C.W.: Performance improvement in DTNs by packet size optimization. IEEE Trans. Aerosp. Electron. Syst. **51**(4), 2987–3000 (2015)
20. Lent, R.: A cognitive networking technique for LTP segmentation. In: The International Wireless Communications and Mobile Computing Conference, June 2020
21. Burleigh, S., Ramadas, M., Farrell, S.: Licklider Transmission Protocol - motivation. RFC 5325, RFC Editor, September 2008
22. Ramadas, M., Burleigh, S., Farrell, S.: Licklider Transmission Protocol - specification. RFC 5326, RFC Editor, September 2008
23. Farrell, S., Ramadas, M., Burleigh, S.: Licklider Transmission Protocol - security extensions. RFC 5327, RFC Editor, September 2008
24. Lent, R.: Regulating the block loss ratio of the Licklider Transmission Protocol. In: 2018 IEEE 23rd International Workshop on Computer Aided Modeling and Design of Communication Links and Networks (CAMAD), pp. 1–6, September 2018
25. Lent, R.: Analysis of bundle throughput over LTP. In: 2018 IEEE 43rd Conference on Local Computer Networks (LCN). pp. 271–274 (October 2018)

Detection of NAT64/DNS64 by SRV Records: Detection Using Global DNS Tree in the World Beyond Plain-Text DNS

Martin Hunek(✉) and Zdenek Pliva

Institute of Information Technology and Electronics, Technical University of Liberec, Studentska 1402/2, 46117 Liberec 1, Czech Republic
{martin.hunek,zdenek.pliva}@tul.cz
https://www.tul.cz

Abstract. Since it has been introduced the NAT64/DNS64 transition mechanism has reputation of method which simply works. This could change as currently used detection method, RFC7050 [16], for this transition mechanism doesn't work with third party/foreign DNS resolvers. These resolvers have been lately introduced by Mozilla Firefox [1] with implementation of DNS over HTTPS. This paper describes problems connected with default usage of third party DNS resolvers and provides a way how to solve issues of RFC7050 [16] with and without third party resolvers.

Keywords: NAT64/DNS64 · DNS · DNSSEC · DoH · RFC7050

1 Introduction

The Internet Protocol version 6 (IPv6) and a whole internet changed a lot during more than 20 years the first standard of IPv6 has been around. At a beginning every device used EUI-64 as its identifier, device autoconfiguration has been split into two services - neighbor discovery for routing and addressing in case of Stateless Address Autoconfiguration (SLAAC) and Dynamic Host Configuration Protocol version 6 (DHCPv6) for everything else. Global Content Distribution Networks (CDNs) didn't exist, cloud was still just a condensed water in the atmosphere and most of the internet services where self-hosted. Because of that the internet itself was truly decentralized network, made of smaller networks connected in several Internet Exchange Points (IXPs).

When going bit more to the history, to the beginning of internet itself, to days when Internet Protocol version 4 (IPv4) switchover happened. The internet was a network of mutually trusted networks with network administrators, who knew each other. And from this age some of the essential protocols have been established. These include telnet, Simple Mail Transfer Protocol (SMTP), Domain Name System (DNS) and many more. Some of them are not in use

© Springer Nature Switzerland AG 2020
P. Gaj et al. (Eds.): CN 2020, CCIS 1231, pp. 27–40, 2020.
https://doi.org/10.1007/978-3-030-50719-0_3

nowadays, some really shouldn't but still are (telnet) and some still performs its essential role in the internet - like DNS.

Unfortunately, we are not living in the same world and we are not using the same internet anymore. Internet has become a network of content providers making their own CDNs distributed around the globe, slowly taking more and more services concentrated into them. Network administrators no longer knows each other, they hardly know a few people from their IXPs. The Internet is no longer just a network of computers. There are tons of devices which are poorly developed/managed and shouldn't be really connected to public network, the internet has become. It is no longer safe trusted network of professionally maintained computers, it is a jungle for masses.

This change in way how the internet is perceived of course initiated changes in some of these protocols, but those changes also caused some collateral damage which would be described in this article.

2 Encrypted Domain Name System Protocols

As already mentioned, one of the most important protocols of the internet, which wasn't originally designed with any security concerns is a DNS. It was introduced as a replacement of distributed host file. However as network protocol it does not provide a same security level as locally managed file and in the same time a DNS itself does not provide any protection against spoofing, other then race condition.

DNS is plain-text protocol without cryptography signatures and without end to end encryption. At the beginning of the internet it really didn't caused an issues as it has been viewed as trusted network. But as the internet became more broadly adopted, it has been realized that this nature of DNS could be leveraged to perform Man in the Middle (MitM) attacks.

In this attack a client is given spoofed replies to its DNS queries. Attacker utilize either closer proximity to a client or simpler and faster DNS software, as it must provide a same reply on every query. As the only first reply received by a client is used and as there is not cryptography signature present in a reply, client has no means to validate received data, so it is inevitably redirected to attacker.

As a DNS doesn't provide any defense against these kind of attacks, a cryptographic signatures have been introduced into DNS tree by Domain Name System Security (DNSSEC). By establishing the chain of trust from IANA maintained root zone up to every single record in every signed zone, the MitM attack has been mitigated. When a zone for which a validating client is performing a query is signed, any manipulation with reply would be detected and rejected as forged. This is inherently safer then connecting to attacker's device. In fact for validating client, a MitM attack changes to Denial of Service (DoS) type of attack, as when a client receives forged reply first and legitimate second, it may ignore both forged and legitimate. The first one would be ignored due to failure in DNSSEC validation and the second one would get ignored as there was already reply received for the same query. For this reason a service would not be accessible so it would mean successful DoS attack[1]. However forged reply must not been cached in client so it would mean when attacker would either stop transmitting packet to client or when legitimate reply would arrive first, then a service would become accessible again.

There are also legitimate reasons to modify DNS replies like Network Address Translation 6-to-4 (NAT64)/Domain Name System 6-to-4 (DNS64) transition mechanism (defined in RFC6146 [14] and further explained in [2]) which had to deal with presence of DNSSEC. This will be mentioned further in this article.

The MitM and DoS are not the only risks connected to plain-text DNS. Essentially when using unencrypted channel over insecure network there is always risk of interception and with that interconnected risk of leakage sensitive information, as well as targeted DoS attacks often also used for government censorship.

This privacy related concerns lead to two independent standards of encrypted DNS protocols. Both of them are using the same principle of encapsulation DNS traffic inside encrypted channel but differs in transport channel.

2.1 DNS-Over-TLS

The first method of encapsulation of DNS traffic is the DNS over TLS (DoT) (RFC7858 [9]). It is actually not more then its name says. It is the plain-text DNS encapsulated in either TLS tunnel in case of Transmission Control Protocol (TCP) transport or the DTLS in case of User Datagram Protocol (UDP) transport.

It uses a separate port number 853, and it is an alternative way of transport for the same servers as used for plain-text DNS on port 53. When a client supports this method of transport then it tries to establish connection on port 853. When server supports DoT, then connection is made and DoT is used for transporting DNS queries.

As a DoT uses the same resolvers as regular DNS, it does not require any detection method other then trying to establish a connection on DoT port. Also by using the same resolvers it does not introduce third parties into communication of a client.

The down side of this method is that it allows easy way how to block client access to DoT by setting a firewall forward chain anywhere between client and resolver blocking port 853 which is used solely by DoT. This would also make government censorship easier.

2.2 DNS-Over-HTTPS

The second method, defined by RFC8484 [8] uses, as the name implies, HTTPS as means of transport.

The main advantage of this method is that it is using well-known port of 443, this makes it quite hard to filter DNS traffic from regular HTTPS content.

---

[1] This behavior can be mitigated by stub resolver and depends on its implementation.

This way it would require performing deep packet inspection and SSL stripping to inspect and filter DNS traffic.

The main disadvantage is a missing detection method. As regular DNS uses IP addresses for accessing resolver, the IP have to use a whole URL to access DoH API. This would seems as a small and insignificant change, however CAs are not issuing TLS certificates for IP addresses and single IP address can host a multiple sites by leveraging SNI. Also URL suffix of DoH API could differ from typical location and in such case using just IP would not provide sufficient information to successful setup DoH.

By missing detection method a DoH clients had to depend either on list of DoH provider shipped with software, or on manual entry made by user. This is especially concerning in case of Mozilla Firefox. Mozilla has made a DoH as a default (currently US only) [1], with default provider being Cloudflare, but with possibility to change to other DoH providers or disabling DoH all together. It is also worth mentioning that one of the DoH providers is also Google (not listed in Mozilla Firefox) and the people working for Google were one which proposed DoH to IETF.

When looking into this from privacy point of view, this leads to less privacy then plain-text DNS. This is caused by introducing a third party into resolving process, in difference to regular DNS, which utilizes Internet Service Provider (ISP) resolver with cached records. By asking DoH provider instead, client is giving that provider every query, which could be directly connected back to client. And by utilizing cookies and other techniques for user tracking on web, it is not just identified as an IP (possibly shared with multiple users in case of IPv4), it is identified as single browser on client.

It is not surprising that people working for Google would propose standard, which would give a third party (DoH provider) all DNS queries, which it would not otherwise get and when it just happen that one of DoH providers is Google itself.

Privacy issues are not the only one connected with DoH. Introducing a third party into DNS also breaks policy based DNS and also split view in DNS. This then breaks current DNS based method for detecting transition mechanisms like NAT64/DNS64 and also situation when network uses private addresses for locally hosted services.

The Fig. 1 shows both traditional DNS (solid line) and DoH route how a DNS query is send and reply is received. It could be seen that when DoH is used every query is send directly to DoH provider and from its web server is processed via traditional DNS. When would a client ask for locally hosted service, then query would be also send via DoH to the DoH provider, which would then query required record from ISP DNS. But this query would have come from WAN facing interface (from public address not belonging to ISP) so outside view would be used. This could potentially lead to unreachable service as ISP could use RFC1918 [15] addresses for local access or requested service could be accessible for locally connected clients only. If so, then client querying record for such service could receive either public address of such service (which would not be reachable by client) or client would receive empty reply as a service would be internal only and it would appear that client is connecting from outside of local network. This is why a DoH breaks split view zones.

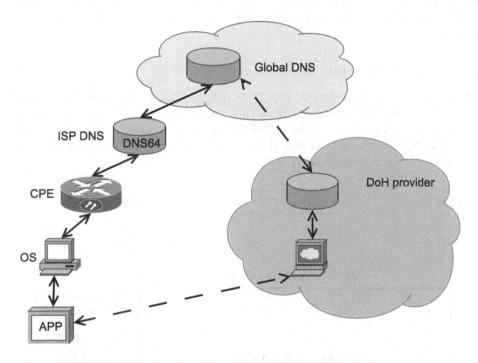

Fig. 1. Comparison of DoH and traditional DNS route of queries

It is fair to say, that both privacy and split view problem is not direct property of DoH. These issues are caused by its implementation and by lack of detection method. There are some studies which are advocating use of DoH like [3] or [13] but such studies are concentrated on performance, availability measurements and stub resolver to recursive resolver link security. From that point of view DoH seems fine, but they are missing problem of metadata leakage to the third party - DoH provider. Such problem is stated in [7].

3   NAT64/DNS64

The NAT64/DNS64 is one of transition mechanisms which utilizes two separate technologies. One is NAT64 which translates between IPv6 and IPv4 and vice versa (this is a difference between NAT44 - commonly referred as Network Address Translation (NAT) and NAT66 which translates addresses inside a same address family).

The second part of this mechanism is DNS64. It provides synthesized AAAA record for services only having A record. This way, as IPv6 has priority over

IPv4, client has possibility to connect to IPv4 only service via IPv6 network. As DNS64 is synthesizing records not originally present in a zone, it conflicts with DNSSEC. This conflict has to be resolved by NAT64/DNS64 detection method, otherwise DNS64 response would get discarded by client so that NAT64 would not be used and if client has no IPv4 connectivity it would not be able to reach IPv4 only service.

3.1 Current Detection Method - RFC7050

Detection method specified in RFC7050 [16] is DNS based solution. It uses a AAAA query for Well-Known IPv4-only Name (WKN) *ipv4only.arpa.* which has got only IPv4 address in the global DNS tree. This query is performed with DNS flag "CD" set to zero so that DNS64 could perform address synthesis. This query and reply is shown by Fig. 2.

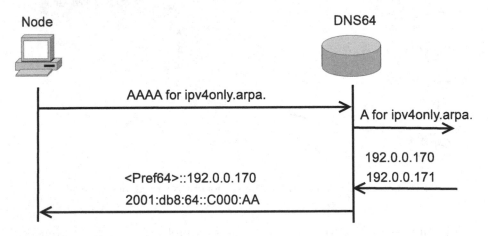

Fig. 2. Detection of NAT64 prefix according to RFC7050 [16] (Source: [10])

When client receives reply containing IPv6 addresses, then DNS64 is present in network and prefixes of these addresses are equal to prefixes used for NAT64. In the example shown in Fig. 2 the NAT64 prefix is *2001:db8:64::/96*.

When there is no DNS64 service provided, then client should receive NODATA[2] status code for AAAA *ipv4only.arpa.* query. The RFC7050 [16] also allows reply of NXDOMAIN for negative answer, however this is against specification of RFC1035 as there is an A record for *ipv4only.arpa.* so there is another record type for WKN while NXDOMAIN would indicate that WKN does not exist in DNS tree which is obviously not true. This means either configuration error, bug in DNS resolver implementation or the *arpa.* zone is not being correctly resolved.

---

[2] NODATA is not actually transmitted as a return code. It is a combination of NOERROR code and missing answer section.

If client is not performing validation of received NAT64 prefix it is allowed to finish detection process in the first step. When client is capable of validation it should proceed with sending PTR queries for every received address and then for every PTR reply it should query AAAA record. Reply of AAAA query then must match with reply for AAAA record of WKN.

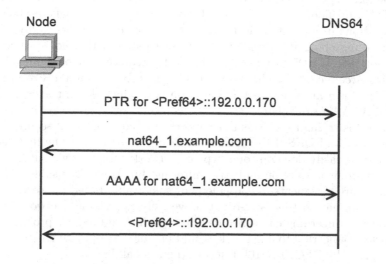

Fig. 3. Validation of NAT64 prefix according to RFC7050 [16] (Source: [10])

Figure 3 shows the whole process of NAT64 prefix validation. It uses generic $<Pref64>::192.0.0.170$ and generic domain *example.com.* but it could be substituted by address from previous example in Fig. 2, in which case address in PTR query and AAAA reply would be *2001:db8:64::c000:aa.* Domain name in PTR reply and AAAA query could be any valid domain name, but it has to be under the ISP control and must lead back to address detected in the first step, otherwise validation would fail.

Nowadays, the most fundamental problem connected with this method would be its dependence on DNS server provided by ISP. Before introduction of DNS the usage of third party DNS servers was a rare setup, which had to be configured manually. Usually a client is provided with DNS resolver address is via autoconfiguration (DHCP or SLAAC) and such information would be also passed to downstream interface of any router, which could be located between a client and ISP autoconfiguration server.

However, after DoH has been introduced and actively used, the third party DNS resolvers have been automatically configured inside a client. Then both system resolver containing NAT64/DNS64 detection method and DNS resolver provided by ISP, which runs DNS64 service, are getting automatically bypassed (shown by Fig. 1). This effectively means no NAT64/DNS64 for clients using DoH and so no IPv4 service in IPv6-only networks which utilizes this transition mechanism.

There are also some security implications connected with this method. The first step the detection process could not be validated by DNSSEC. This is due to the fact that DNS64 provider (usually ISP), is not legitimate holder of *arpa.* zone. This means that AAAA record for WKN could not be signed and it has to be solved by setting a "CD" flag as mentioned previously. When this first query would be intercepted and replied with forged records it would allow to perform DoS, MitM and flooding attacks. Validation would not necessary protect from these situations because it could be successfully passed for every address which would have matching PTR and AAAA record pair. Furthermore, standard is using word "SHOULD" instead of "MUST" for requirement of signing NAT64 AAAA resource record and there is no requirement for signing PTR record. This allows to perform mentioned attacks even on host which doesn't have matching PTR and AAAA record pair.

Standard is trying to address this concerns by requirement of secure channel between client and DNS64 server and via secure domain list. The first requirement could be easily done on some type of network - fixed service utilizing star topology with encapsulation and strict filtering; but is hardly usable for others - utilizing shared segments, bus topology or some radio based networks.

Second mitigation tool is also not universally applicable. Networks forcing strict policies concerning connected devices, which would require provisioning of device prior connection to a network, would be able to populate trusted domain list as required by RFC7050 [16]. But for others, small ISPs using stock firmware for CPEs or on networks with BYOD policy, this requirement would not be possible to fulfill.

3.2   Alternative Means of Detection

Following methods have one thing in common, there are not widely used. It is either because they are using not widely adopted protocol or because of they haven't been standardized yet.

The most relevant of these method already standardized is RFC8115 [5]. It uses DHCPv6 option code 113 called "OPTION_V6_PREFIX64". This option includes two multicast prefixes and prefix lengths and single unicast NAT64 prefix. Problem of this method is that it uses DHCPv6 - protocol which is not mandatory and which is not implemented is some clients (Android) and because of that it is not widely deployed in residential networks. This causes so called "circulus vitiosus" as no support in clients means no deployment in a networks and no deployment means no need for implementation in clients. This religious battle of Android against DHCPv6 caught RFC8115 [5] in it, which is a shame as it could work quite well, maybe even with DoH.

Another possibly relevant way how to detect NAT64/DNS64 is via Port Control Protocol (PCP) - specified by RFC7225 [4]. The PCP is protocol for controlling behavior of NAT and firewall by a client requests. The RFC7225 [4] is extending PCP by option with code 129, which includes NAT64 prefix and its length as well as IPv4 prefixes used for translation (optionally). Even that PCP showed an interesting concept of client managing firewall and NAT of upstream router, it has not been adopted inside either residential or enterprise networks. Also as non-essential protocol for autoconfiguration, even knowledge of its existence is pretty low among network administrators.

There is also one new method of NAT64/DNS64 detection [6], which uses ICMPv6 Router Advertisement extension. Same as previous method it includes NAT64 prefix, its encoded length and it adds encoded validity time. However it differs in protocol used, which in this case is essential for autoconfiguration of any IPv6 client and further more the ICMPv6 is essential for IPv6 itself. This is a huge advantage of this method and if it gets standardized it has a potential to solve current issues of NAT64/DNS64 detection. Only possible pitfall could be hidden within operating systems (network stack) as it has to provide a way how to distribute learned NAT64 prefix to application including DNS resolver (web browser in case of DoH). If there would not be such interface, then DoH enabled application would have to implement ICMPv6 Neighbor Discovery protocol by themselves in order to support this method.

4   Proposed Detection Method

4.1   Reasoning

Due to the standardization of DoH and its introduction of third party DNS providers as a default settings the problems of RFC7050 [16] had to be addressed. But because failure of previous standards in real network adoption, we had to introduce some design goal for new method.

These design goals are:

Goal 1 No new protocol or alteration of existing one.
Goal 2 Utilize widely supported protocols.
Goal 3 Utilize information already provided by network.
Goal 4 Must work with foreign DNS.
Goal 5 Must not require DNS64 synthesis on a host.
Goal 6 Must not require prior provisioning.
Goal 7 Must provide secure detection over insecure channel.

Goal 1 is purely motivated by ease of standardization process - less changes into working and already deployed protocols means less testing and lower probability of introducing vulnerabilities. The second goal is motivated by situation of RFC7225 [4] and RFC8115 [5]. Both are using non-essential protocols, so any deployment of these standards requires configuration work done by network administrator. This is strongly connected with design goals 3 and 6 as if deployment of detection method is easier for network administrator it is more likely to be deployed.

The fourth design goal is motivated by the current problem of RFC7050 [16] with DoH. The fifth goal is reaction to method currently discussed on 6man work group which utilizes Router Advertisement messages. This method does not provide detection of DNS64, just NAT64 and client is then forced to perform DNS64 itself. It is certainly a way how to solve DNS64 problem but this solution adds requirements on client implementation which hasn't been there up to this point. This could potentially slow down deployment process or restrict its audience on devices with limited resources.

The final design goal is also reaction on RFC7050 [16] and its requirements of secure channel. By somehow expecting trustworthy network, method usage would be either limited or the worse scenario - method would introduce vulnerability to a network, which would deploy it regardless of its prerequisites.

4.2   Principle of Method

The newly proposed method [11] utilize DNS, similar way how it is done in RFC7050 [16], but instead of using split view to *arpa.* zone and well-known name *ipv4only.arpa.* it utilizes SRV records inside local domain. These records are globally resolvable and because of that, they can be used even in the situations when client is using third party DNS resolver.

Draft of this method [11] introduces two new services - SRV records: One for signaling NAT64 prefix (and optionally also outside IPv4 addresses) a the second for providing address of DNS64 resolver. Example of such records in operator zone is shown in Listing 1.1.

```
$ORIGIN example.net

% NAT64 records
_nat64._ipv6    IN SRV  5 10 9632 nat64-pool.example.net
nat64-pool      IN AAAA 2001:db8:64:ff9b::c000:aa
nat64-pool      IN A    192.0.2.64

% DNS64 records - stating priorities
_dns64._tcp     IN SRV  5   10 53   dns64.example.net
_dns64._udp     IN SRV  10  10 53   dns64.example.net
_dns64._tcp     IN SRV  20  10 443  dns64.example.net
dns64           IN AAAA 2001:db8::53
```

A zone of this example contains single NAT64 prefix (*2001:db8:64:ff9b::/96*) which is translated into single IPv4 address (*192.0.2.64*). Length of both prefixes are encoded into port number field of SRV record by prepending an IPv6 prefix length before an IPv4 prefix length. This field could be safely used as IP doesn't have port numbers (port number is used in layer 4 protocols - TCP, UDP etc.) and as 16b number it can handle theoretical maximum of /128 IPv6 prefix length and /32 of IPv4 which would make 12832 in decimal notation. This information is only optional as it is indicating size of IPv4 pool used for translation. As long as operator is using typical prefix length of /96 for embedding an IPv4 address it could just set port number field to 0. This would be signal to client that information about pool size is not available and that IPv4 address is just appended after a NAT64 prefix. Otherwise format of this record does not differ from standard SRV record.

The second proposed record represents DNS64 resolvers. In the example it shows a single DNS64 resolver accessible by plain-text DNS protocol over TCP (most preferred) and UDP and by DoH (least preferred). Implementation of this record is not strictly required for this method to work, but it adds an interesting possibilities for network configuration. Because of this record, operator is given a tool how to indicate client on which servers the DNS64 service is provided and which protocols are used and preferred. This also adds a possibility to run DNS64 service outside of main recursive DNS infrastructure and also provides an easy way how to provide client domain names of ac:DoH servers (which so far does not exits).

4.3   Client Behavior

Client, when connected to a network, will typically receive some autoconfiguration information. This in IPv6 capable network consist of Router Advertisement message which can optionally include *DNSSL* option (domain search list - RFC8106 [12]), and/or DHCPv6 which can include option codes 24, 39, 57, 74, 118 and others which can provide local domain as well. Detection of local domain is important for this proposed method to work, as it uses DNS queries for getting SRV records in this local domain.

When this information is not provided passively as part of autoconfiguration process, client can perform PTR query for its own IPv6 address. This should produce dynamically generated record pointing to resource holder domain (typically either ISP or large company). Such domain can be then used to perform detection process. The last usable source of an information about local domain could be client's Fully Qualified Domain Name (FQDN), which should also include local domain.

After local domain is established, client can start detection method. This consist of querying *_nat64._ipv6* subdomain for every local domain candidate, and if it is not capable of DNS64 synthesis, it should also query for DNS64 service record (*_dns64*) under both TCP and UDP. All replies for this queries must be signed by DNSSEC and their signatures must be valid. If client is capable of validating DNSSEC signatures, it must discard any record with invalid or missing signature. This requirement solves design goal number 7 as when network utilizes countermeasures against spoofing RA or DHCPv6 (RA-guard and DHCPv6 snooping), it provides sufficient proof that provided prefixes and DNS64 servers are legitimate (or at least their addresses). This also conform design goal 6 as DNSSEC key of root zone is included in DNS resolver code and no other key has to be distributed.

If client is not capable of performing DNS64 address synthesis, it can use DNS64 servers provided by SRV record. Default behavior of client in this case should be to use these servers only when it receives NODATA reply for AAAA query. This would most likely indicate presence of an A record, so that it could be IPv4-only service and NAT64/DNS64 must be used to access it. Client could potentially use DNS64 servers from SRV record as default, but it could potentially override its user's wishes and lead for unintended leak of sensitive information so it should be uses only when there is no reply on AAAA queries.

If a client is capable of performing DNS64 address synthesis, then it can do that with received NAT64 prefix or it can use DNS64 servers. This should be configurable but either option would work.

4.4   Fulfillment of Design Goals

Proposed method fulfills the goal 1 as it is not introducing any new protocols, and it fulfills goal 2 as it is utilizing DNS as source of configuration information and DNS is essential network protocol. Further more, if there will any future version of DNS transport, this method should work with it as it is independent of transport method.

Design goal 3 is at least partially fulfilled. As this method uses *DNSSL* as source of information about local domain (this would be usually present in the network) or DHCPv6 options (which might not present already). In case of simple enterprise network without routers managed by end user (CPE), network administrator can just utilize *DNSSL* already configured and just add corresponding NAT64 SRV and AAAA records. This can be done in single place - forward zone file. When there are CPEs in the network then network administrator must also deploy DHCPv6 option preferably code 57 or dynamic creation of PTR record for client addresses. That is why this goal is considered fulfilled partially - it can work in some networks out of the box, but in other networks additional steps must be taken.

Goal 4 is fulfilled as method uses global DNS tree, so it would work with any recursive resolver and with any transport method. Goal 5 is fulfilled with providing DNS64 SRV records. Goals 6 and 7 are fulfilled with strict requirement of DNSSEC. Only security measures required for network is basic security of autoconfiguration methods, but in absence of these security measures it would not add any other issues to those, which would be already present in such network. Without secure channel it is still possible to perform some attacks on this method like DoS or to intercept DNS queries between client and resolver, but with DNSSEC it is not possible to inject client with rogue prefix as long as client is validating responses.

5   Conclusion

The NAT64/DNS64 is really easy to use and reliable transition mechanism, so it would be a shame to loose such possibility to widely deploy it to phase out the IPv4 in access networks. But with current detection methods this could become a reality, as the most deployed detection method (RFC7050 [16]) is not compatible with DoH resolver in Mozilla Firefox [1]. There are also other methods, how to provide detection of NAT64 prefix, but these either depends on non-essential protocols (not necessary deployed in access networks) or they are not yet standardized.

Detection method described in this document proposes one possible way how to solve issues related to RFC7050 [16] and how to move an information about both NAT64 and DNS64 from local view of *arpa.* zone to operator's global zone. This also allows to secure all related records by DNSSEC, without need to disable validation in any point so that detection can be done over insecure channel.

Acknowledgement. This work was supported by the Student Grant Scheme at the Technical University of Liberec through project nr. SGS-2019-3017.

References

1. Firefox DNS-over-HTTPS (2019). https://support.mozilla.org/en-US/kb/firefox-dns-over-https
2. Bagnulo, M., Garcia-Martinez, A., Beijnum, I.V.: The NAT64/DNS64 tool suite for IPV6 transition. IEEE Commun. Mag. **50**(7), 177–183 (2012). https://doi.org/10.1109/MCOM.2012.6231295
3. Boettger, T., et al.: An empirical study of the cost of DNS-over-HTTPS. In: ACM Internet Measurement Conference (IMC) (2019)
4. Boucadair, M.: Discovering NAT64 IPv6 prefixes using the Port Control Protocol (PCP). RFC 7225, May 2014. https://doi.org/10.17487/RFC7225. https://rfc-editor.org/rfc/rfc7225.txt
5. Boucadair, M., Qin, J., Tsou, T., Deng, X.: DHCPv6 option for IPv4-embedded multicast and unicast IPv6 prefixes. RFC 8115, March 2017. https://doi.org/10.17487/RFC8115. https://rfc-editor.org/rfc/rfc8115.txt
6. Colitti, L., Linkova, J.: Discovering PREF64 in router advertisements. Internet-Draft draft-ietf-6man-ra-PREF64-09, Internet engineering task force, December 2019. https://datatracker.ietf.org/doc/html/draft-ietf-6man-ra-pref64-09. (Work in progress)
7. Hoang, N.P., Lin, I., Ghavamnia, S., Polychronakis, M.: K-resolver: towards decentralizing encrypted DNS resolution. In: The NDSS Workshop on Measurements, Attacks, and Defenses for the Web 2020 (MADWeb 2020), pp. 1–7, February 2020. https://doi.org/10.14722/madweb.2020.23009
8. Hoffman, P.E., McManus, P.: DNS queries over HTTPS (DoH). RFC 8484, October 2018. https://doi.org/10.17487/RFC8484. https://rfc-editor.org/rfc/rfc8484.txt
9. Hu, Z., Zhu, L., Heidemann, J., Mankin, A., Wessels, D., Hoffman, P.E.: Specification for DNS over Transport Layer Security (TLS). RFC 7858, May 2016. https://doi.org/10.17487/RFC7858. https://rfc-editor.org/rfc/rfc7858.txt
10. Hunek, M., Pliva, Z.: DNSSEC in the networks with a NAT64/DNS64. In: 2018 International Conference on Applied Electronics (AE), pp. 1–4, September 2018. https://doi.org/10.23919/AE.2018.8501446
11. Hunek, M.: NAT64/DNS64 detection via SRV records. Internet-Draft draft-ietf-v6ops-nat64-srv-00, Internet Engineering Task Force, March 2019. https://datatracker.ietf.org/doc/html/draft-ietf-v6ops-nat64-srv-00. Work in Progress
12. Jeong, J.P., Park, S.D., Beloeil, L., Madanapalli, S.: IPv6 router advertisement options for DNS configuration. RFC 8106, March 2017. https://doi.org/10.17487/RFC8106. https://rfc-editor.org/rfc/rfc8106.txt
13. Lu, C., et al.: An end-to-end, large-scale measurement of DNS-over-encryption: how far have we come? In: Proceedings of the Internet Measurement Conference IMC 2019, pp. 22–35. Association for Computing Machinery, New York (2019). https://doi.org/10.1145/3355369.3355580

14. Matthews, P., van Beijnum, I., Bagnulo, M.: Stateful NAT64: Network Address and Protocol Translation from IPv6 Clients to IPv4 Servers. RFC 6146, April 2011. https://doi.org/10.17487/RFC6146. https://rfc-editor.org/rfc/rfc6146.txt
15. Moskowitz, R., Karrenberg, D., Rekhter, Y., Lear, E., de Groot, G.J.: Address allocation for private Internets. RFC 1918, February 1996. https://doi.org/10.17487/RFC1918. https://rfc-editor.org/rfc/rfc1918.txt
16. Savolainen, T., Korhonen, J., Wing, D.: Discovery of the IPv6 prefix used for IPv6 address synthesis. RFC 7050, November 2013. https://doi.org/10.17487/RFC7050. https://rfc-editor.org/rfc/rfc7050.txt

Quantum Router for Qutrit Networks

Marek Sawerwain[1(✉)] and Joanna Wiśniewska[2]

[1] Institute of Control and Computation Engineering, University of Zielona Góra,
Licealna 9, 65-417 Zielona Góra, Poland
M.Sawerwain@issi.uz.zgora.pl
[2] Institute of Information Systems, Faculty of Cybernetics, Military University
of Technology, Gen. S. Kaliskiego 2, 00-908 Warsaw, Poland
jwisniewska@wat.edu.pl

Abstract. Networks of quantum circuits or, more generally, networks transmitting quantum information will need, just like classical networks (e.g. internet), a mechanism for directing data to adequate nodes. Routing, understood as packet switching, is one of the most important processes in classical networks. The issue of routing is also present in quantum networks and an appropriate construction of a quantum router is required to transfer data to specific points in the network. We describe an implementation of a router for qutrits in this chapter. The router is four-qutrit quantum circuit (with one controlling unit). The efficiency and the accuracy of router's work is tested by the Fidelity measure. The circuit's dynamics is expressed by a Hamiltonian where the role of generalized Pauli operators is played by the Gell-Mann operators.

Keywords: Quantum networks · Quantum router · Qutrits

1 Introduction

Transferring information is not the only role of networks. They may be seen as something more, e.g. tensor networks [11], neural networks [19] or quantum circuits [22]. Undoubtedly, processing and transferring of information in quantum networks is a problem which should be solved to efficiently realize quantum computations [30,31], and utilize quantum communication protocols. It should be emphasized that, nowadays, quantum networks [27] are intensively studied, and many tools are constructed to investigate behavior of these structures, like quantum networks – or even quantum internet – simulators [8–10].

Presently, the transfer of quantum data is based on quantum spin-chains [5, 21], and entangled qubits [20,23,29]. Different physical elements are considered as components of future quantum networks. Many elements of classical networks, like switches, repeaters, and routers, have their quantum equivalents [1–4,14,28].

It should be emphasized that mentioned components of networks, especially routers, are not only discussed as theoretical devices. We can find their experimental physical implementations, for example with the use of coupled harmonic system [25], quantum tunneling effect [18], or superconducting circuits [7,26].

© Springer Nature Switzerland AG 2020
P. Gaj et al. (Eds.): CN 2020, CCIS 1231, pp. 41–51, 2020.
https://doi.org/10.1007/978-3-030-50719-0_4

In this work, we would like to show that a quantum router may be implemented for higher units of quantum information, i.e. qutrits. We describe the basis definition of a router, and its dynamics as a Hamiltonian where, because of qutrits, Pauli operators are substituted by the Gell-Mann matrices. The given router definition is consistent with the form of the qubit router, presented in professional literature, so it may be treated as a step in a direction of generalization to the qudit router. We present the values of Fidelity measure which proof that the proposed Hamiltonian correctly realizes tasks of the qutrit router.

The paper is organized as follows. In Sect. 2, we describe the basic information concerning qudits, and qutrits in particular. We adduce also the Gell-Mann operators which are required in the Hamiltonian construction (the Hamiltonian describes the dynamics of router's operating).

Section 3 contains the idea and definition of the qutrit router. The router is presented as a quantum circuit, and, what is more important, as the Hamiltonian. Experiments inspecting the router's operating are shown in Sect. 4. There, the values of Fidelity measure for the routing process are presented. The summary is positioned in Sect. 5. Acknowledgments and references section end the article.

2 Preliminaries

Both, in classical and quantum computing, the definition of a unit of information is required. The construction of presently used computers impose a bit as the basic unit. Naturally, the first algorithms for quantum computers were also proposed for quantum bits, so-called qubits. However, the technical development enables the utilization of quantum information units with a freedom level greater than two (higher freedom level causes higher informational content what allows obtaining the result of computation with fewer operations).

Let us define a qudit as a general unit of quantum information with the freedom level $d \geq 2$. A quantum state of a single qudit may be expressed as a normalized d-entity column vector. We denote this kind of vectors, in the Dirac notation, as $|\cdot\rangle$, e.g.

$$|\psi\rangle = \begin{pmatrix} \alpha_0 \\ \alpha_1 \\ \dots \\ \alpha_{d-1} \end{pmatrix}, \tag{1}$$

where the normalization condition requires $\sum_{i=0}^{d-1} |\alpha_i|^2 = 1$, and α_i are the complex numbers. In next sections of the text, we denote a quantum state also as ψ what still means the correct state in the Dirac notation.

If the quantum state is created by more than one qudit, its states' vector is calculated as a tensor product of all one-qudit states vectors. For example, let us have two qudits: $|\psi\rangle, |\phi\rangle$, with different freedom levels: a and b, respectively. The state of these qudits, joined in one quantum register, is:

$$|\Psi\rangle = |\psi\rangle \otimes |\phi\rangle, \tag{2}$$

where the dimensionality of a vector $|\Psi\rangle$ is equal to $a \cdot b$ (the dot symbol represents the scalar product of two numbers). Of course, the joined qudits may have the same freedom levels. The symbol of tensor product is usually omitted, so the above state $|\Psi\rangle$ may be written as $|\psi\phi\rangle$.

Quantum states may be also described by the superposition equation. In this case, we need to define a concept of a computational basis. Just like in positional number system theory, we need to clearly point out a representative for each accepted value $i = 1 \ldots d$. In quantum computing these values are substituted by vectors. The computational basis for a d-level single qudit contains d orthonormal vectors (the orthogonality ensures a possibility to distinguish the elements, and the normality guaranties obtaining a correct quantum state). The most popular computational basis is so-called standard basis. Vectors in this basis have one element equal to 1, and other $(d - 1)$ elements equal 0. Of course, in each vector the non-zero element occupies different position:

$$|0\rangle = \begin{pmatrix} 1 \\ 0 \\ \cdots \\ 0 \end{pmatrix}, \ |1\rangle = \begin{pmatrix} 0 \\ 1 \\ \cdots \\ 0 \end{pmatrix}, \ \ldots, \ |d-1\rangle = \begin{pmatrix} 0 \\ 0 \\ \cdots \\ 1 \end{pmatrix}. \tag{3}$$

The superposition is one of the characteristic features of quantum states. The superposition equation shows that a quantum state may be a mixture of basis states with the proportions described by probability amplitudes α_i:

$$|\psi\rangle = \alpha_0|0\rangle + \alpha_1|1\rangle + \ldots + \alpha_{d-1}|d-1\rangle, \tag{4}$$

where $\sum_{i=0}^{d-1} |\alpha_i|^2 = 1$, and α_i are the complex numbers.

To realize the computation on quantum states, we need operators. These operators may be expressed as unitary matrices sized $d \times d$, if they act on single qudit with the freedom level d. If the state contains n qudits (all with the same freedom level), the size of an operator's matrix representation is $d^n \times d^n$, because matrices affecting sequent qudits are tensor multiplied just like in Eq. (2). Of course, if we do not want to change the state of one (or more) particular qudit in the register, we can use the identity matrix $I_{d \times d}$ in the tensor multiplication.

In this work, we describe a router acting on qudits with the freedom level $d = 3$ – called qutrits. Now, we would like to present basic quantum gates, but with the restriction to qutrit gates.

The fundamental rotations which may be realized on one qutrit are given by the Gell-Mann matrices:

$$\lambda_1 = \begin{pmatrix} 0 & 1 & 0 \\ 1 & 0 & 0 \\ 0 & 0 & 0 \end{pmatrix}, \ \lambda_2 = \begin{pmatrix} 0 & -i & 0 \\ i & 0 & 0 \\ 0 & 0 & 0 \end{pmatrix}, \ \lambda_3 = \begin{pmatrix} 1 & 0 & 0 \\ 0 & -1 & 0 \\ 0 & 0 & 0 \end{pmatrix}$$

$$\lambda_4 = \begin{pmatrix} 0 & 0 & 1 \\ 0 & 0 & 0 \\ 1 & 0 & 0 \end{pmatrix}, \ \lambda_5 = \begin{pmatrix} 0 & 0 & -i \\ 0 & 0 & 0 \\ i & 0 & 0 \end{pmatrix}, \ \lambda_6 = \begin{pmatrix} 0 & 0 & 0 \\ 0 & 0 & 1 \\ 0 & 1 & 0 \end{pmatrix} \tag{5}$$

$$\lambda_7 = \begin{pmatrix} 0 & 0 & 0 \\ 0 & 0 & -i \\ 0 & i & 0 \end{pmatrix}, \ \lambda_8 = \frac{1}{\sqrt{3}} \begin{pmatrix} 1 & 0 & 0 \\ 0 & 1 & 0 \\ 0 & 0 & -2 \end{pmatrix}$$

The above operators may be utilized to construct unitary counterparts of Pauli gates: X, Y, Z. The set of generalized operators contains more elements, e.g. there are three equivalents of the X gate for qutrits:

$$X_1 = e^{\frac{i}{3}\lambda_1}, \ X_2 = e^{\frac{i}{3}\lambda_4}, \ X_3 = e^{\frac{i}{3}\lambda_6}. \tag{6}$$

The counterparts of the Y gate are built with the use of $\lambda_2, \lambda_5, \lambda_7$ operators, and λ_3, λ_8 operators serve to define equivalents of the Z gate.

Another powerful feature of quantum systems, next to the superposition, is an entanglement [12]. This phenomenon is a kind of dependency between quantum states of qudits joined in one register. Colloquially speaking, modifying the state of one qudit (with the use of quantum gate) causes a change of other qudit/qudits which take a part in the entanglement. The entanglement takes place when the state of the register cannot be expressed as a tensor product of all single qudits involved in this system.

3   Quantum Router for Qutrits

In this work, the input qutrit is denoted as $|\psi_I\rangle$, and its state may expressed as:

$$|\psi_I\rangle = \alpha|0\rangle + \beta|1\rangle + \gamma|2\rangle, \tag{7}$$

where $\alpha, \beta, \gamma \in \mathbb{C}$ and $|\alpha|^2 + |\beta|^2 + |\gamma|^2 = 1$. Naturally, the qutrit $|\psi_I\rangle$ is a data input for the router.

The output qutrits (and their states) are described as $|\psi_1\rangle, |\psi_2\rangle, |\psi_3\rangle$ or just $|\psi_1\psi_2\psi_3\rangle$. This three-qutrit register is an output of the router.

There is another qutrit in the router which is a controlling unit – its symbol is $|\psi_C\rangle$, and it accepts exclusively three quantum states: $|0\rangle, |1\rangle, |2\rangle$. The controlling qutrit's state decides about the position of $|\psi_I\rangle$ in the final state of the quantum register. Generally, the state of whole router may be denoted as the register:

$$|\Psi\rangle = |\psi_I\rangle|\psi_1\psi_2\psi_3\rangle|\psi_C\rangle. \tag{8}$$

The way the router operates, for the three fundamental states of controlling qutrit, may expressed as:

$$\begin{aligned}
|\psi_I\rangle|000\rangle|0\rangle &\longrightarrow |0\rangle|\psi_I 00\rangle|0\rangle, \\
|\psi_I\rangle|000\rangle|1\rangle &\longrightarrow |0\rangle|0\psi_I 0\rangle|1\rangle, \\
|\psi_I\rangle|000\rangle|2\rangle &\longrightarrow |0\rangle|00\psi_I\rangle|2\rangle.
\end{aligned} \tag{9}$$

If the controlling qutrit is in the superposition of standard basis states: $|\psi_C\rangle = \alpha|0\rangle + \beta|1\rangle + \gamma|2\rangle$, then the router's construction affects the quantum state as follows:

$$|\psi_I\rangle|000\rangle|\psi_C\rangle \longrightarrow \alpha|0\rangle|\psi_I 00\rangle|0\rangle + \beta|0\rangle|0\psi_I 0\rangle|1\rangle + \gamma|00\psi_I\rangle|2\rangle, \tag{10}$$

and it means the entanglement of the controlling qutrit with the output qutrits.

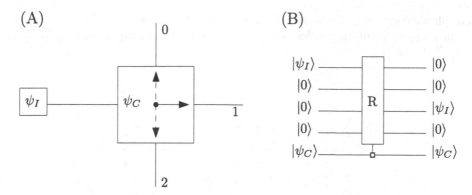

Fig. 1. The operation scheme (A) and the general form of the quantum circuit (B) for the router. Due to operating, one of the output qutrits accepts the state $|\psi_I\rangle$, and other output qutrits are equal $|0\rangle$. The circuit (B) realizes the transfer of information $|\psi_I\rangle$ to the output 1. The unitary operation R symbolizes the router which is controlled by the state of the fifth qutrit $|\psi_C\rangle$ (illustrated as the small empty square)

The description of the router, according to Eq. 9, is depict in Fig. 1. The information written in the state $|\psi_I\rangle$ is routed in a direction defined in $|\psi_C\rangle$ (it appears in one of the outputs: 0, 1 or 2).

It is also interesting to analyze a system built of a few routers. Such a bus of routers allows sending an information to particular nodes in the whole quantum network. Figure 2 depicts exemplary scheme of a five-router bus where the controlling qutrits states $|\psi_{C_0}\psi_{C_1}\psi_{C_2}\ldots\rangle$ point the router's output O_i for the information $|\psi_I\rangle$ to be transferred.

If the qutrit state ψ_I is expected to be routed to the output O_3, the controlling qutrits should be configured: $|111BB\rangle$ (letters B symbolize that qutrits ψ_{C_3} and ψ_{C_4} may accept any basis states – without any influence on the output O_3). The qutrits ψ_{C_3} and ψ_{C_4} are significant for the output O_{10}. To send ψ_I to O_{10}, the controlling qutrits should be $|12B02\rangle$ – as we can see now, the state of $|\psi_{C_2}\rangle$ may be one of three standard basis states, and the ψ_I will be still transferred to O_{10}.

Naturally, the state of controlling qutrits clearly defines the output. If the input information shows up in more than one output, we deal with a phenomenon of entanglement. It is not welcome if we discuss the basic function of the router. On the other hand, we can utilize the entangled states in different outputs as a background in solving other issues in the field of quantum computing.

It should be mentioned that the router transfers information from the input to one of the outputs, and just like for qubits, it is possible to induce entangled states during this process. Naturally, we can build a network of routers, but its structure is a chain or a two-dimensional grid (Fig. 2 depicts such a grid). An analysis of connections between qutrits in multidimensional grids seems very difficult because of the entanglement's presence – there are no methods of entanglement classification, especially for so-called multibody entanglement in real physical systems. There are works concerning routers for multidimensional quantum states, but they refer only to the router's operating on the entangled state [6,13,15,16].

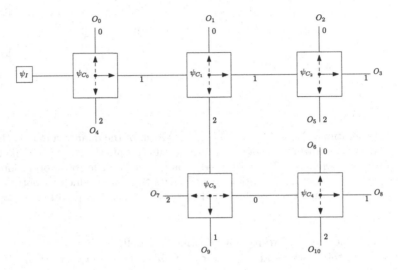

Fig. 2. The exemplary five-router bus. The input state ψ_I may be transferred to one of the outputs O_i, $i = 0, 1, \ldots$, by assigning proper values to controlling states ψ_{C_0}, $\ldots$, ψ_{C_4}. The state of first three qutrits $\psi_{C_0}\psi_{C_1}\psi_{C_2} = |111\rangle$ directs the state ψ_I to the output O_3 (the state of other units has no influence on the system)

In Fig. 1, we can see the controlled unitary operation R which describes the operating of the router. The matrix form of R may be expressed as a permutation operator:

$$
U_0 = \begin{bmatrix}
\bullet & & & & & & & & & \\
 & \bullet & & & & & & 1 & & \\
 & & 1 & & 1 & & & & & \\
 & & 1 & & & 1 & & & & \\
 & & & 1 & & & & 1 & & \\
 & & & 0 & & & 1 & & & \\
 & & & & 0 & & 1 & & & \\
 & 1 & & & & \bullet & & & & \\
 1 & & & & 0 & & & & & \\
 1 & & 1 & & 1 & & & & & \\
 & & 1 & & & 1 & & & & \\
 & & 1 & & & 1 & & & & \\
 & & & & & & \bullet & & \\
 & & & & & & & \bullet & \\
\end{bmatrix},
$$

(11)

because of the system's dimension, we present operator's abbreviated version, including omitted values 1 on the main diagonal and proportion of matrix is not preserved (the dimensions of the full matrix are 243×243).

Of course, the presented matrix is only a part of the permutation matrix, which realizes the task of the router, and it is directly defined, i.e. the digits "1" are placed in the crossings of rows and columns between which the transfer of information should occur.

However, the operator's description given in Eq. 11 does not show the inner actions between qutrits. This kind of insight offers a Hamiltonian (notation $\lambda_7^{(2)}$ means that the operator is used on the second controlling qutrit – the third router's output):

$$
\begin{aligned}
H = &-\frac{1}{2}(\Delta_1 \lambda_3^{(1)} \lambda_8^{(1)} + \Delta_3 \lambda_3^{(2)} \lambda_8^{(2)} + \Delta_3 \lambda_3^{(3)} \lambda_8^{(3)}) \\
&+ J^Z (\lambda_3^{(1)} \lambda_8^{(1)} + \lambda_3^{(2)} \lambda_8^{(2)} + \lambda_3^{(3)} \lambda_8^{(3)})(\lambda_3^{(C)} \lambda_8^{(C)}) \\
&+ \frac{1}{2} J^X \left((\lambda_1^{(I)} \lambda_4^{(I)} \lambda_6^{(I)}) \left(\lambda_1^{(1)} \lambda_4^{(1)} \lambda_6^{(1)} + \lambda_1^{(2)} \lambda_4^{(2)} \lambda_6^{(2)} + \lambda_1^{(3)} \lambda_4^{(3)} \lambda_6^{(3)} \right) \right. \\
&\left. + \lambda_2^{(I)} \lambda_5^{(I)} \lambda_7^{(I)}) \left(\lambda_2^{(1)} \lambda_5^{(1)} \lambda_7^{(1)} + \lambda_2^{(2)} \lambda_5^{(2)} \lambda_7^{(2)} + \lambda_2^{(3)} \lambda_5^{(3)} \lambda_7^{(3)} \right) \right)
\end{aligned}
\tag{12}
$$

The values Δ_1, Δ_2, Δ_3 symbolize frequencies of qutrit transitions between basis states. The frequency J^X denotes the coupling between input qutrit and output qutrits. While, J^Z is the frequency of coupling between output qutrits and controlling unit. Theoretically, these parameters may be selected independently one to another. However, if we want to send the input qutrit state to one of the outputs, the parameters have to meet:

$$
\Delta_1 = -\Delta_2 = \frac{\Delta_3}{2} = 4J^Z.
\tag{13}
$$

Furthermore, we assume that $J^Z > J^X$, and J^Z have to be significantly greater than J^X.

Remark 1. The symbols Δ_i, J^X, J^Z keep their meaning just like for qubits [7]. However, the Pauli operators have to be replaced by the Gell-Mann operators. The given schema may be generalized for qudits, and then $SU(d)$ unitary group operators have to be used [17].

The unitary operator U, describing the router's operating, may be defined with the direct use of H:

$$
U(t) = e^{-i\frac{\pi}{2} t H}
\tag{14}
$$

where $t \in \mathbb{R}$ is the time variable.

4   Numerical Experiments

One of the most important parameters of the router's operating is the accuracy. Of course, presenting quantum operation as the U_0 leads to the perfect results – during a simulation with the use of such a permutation operator, we obtain an output vector as a product of multiplication matrix by vector, and calculated value of Fidelity measure equals one. However, this procedure is purely theoretic. More realistic system's behaviour may be obtained by utilizing a Hamiltonian. Let U be a Hamiltonian-based operator, a value of the Fidelity measure (denoted by the capital letter F) in a moment t is calculated as:

$$F(t) = |\langle\psi_o|U(t)|\psi\rangle|^2, \tag{15}$$

where ψ_o represents the correct final quantum state (after the router's operating), $U(t)|\psi\rangle$ is the router's state for the moment t, if the initial state was ψ. The above definition of the Fidelity measure allow us to evaluate if the whole router works correctly.

Furthermore, in the case analyzed in this paper, it is important to employ the average Fidelity measure (denoted by the letter $\bar{F}$):

$$\bar{F} = \int \langle\psi|\hat{U}^\dagger \mathcal{E}(\psi)\hat{U}|\psi\rangle d\psi. \tag{16}$$

We integrate the area of all input states as a quantum map $\mathcal{E}$ for the router. The operator U denotes the final operation, correctly realizing the router's operating. In our work $\hat{U} = U_0$.

As in [24], the average Fidelity value may be calculated as:

$$\bar{F}(\psi, U_0, M) = \frac{1}{n(n+1)}\left(\text{Tr}\left(MM^\dagger\right) + |\text{Tr}\left(M\right)|^2\right) \quad and \quad M = U_0^\dagger U(t). \tag{17}$$

This way of Fidelity computing does not require the state ψ value. It means that only the forms of U_0 and $U(t)$ influence the value of the average Fidelity measure. The dimension of the state ψ is n.

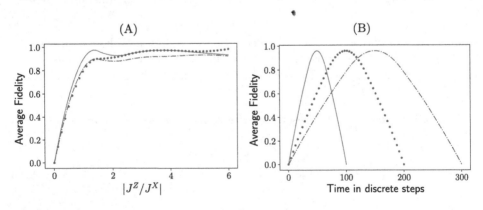

Fig. 3. The changes of average Fidelity value (A) during the router's operating for states $|0\rangle$ (red solid line), $|+\rangle$ (green dotted), $\frac{1}{\sqrt{2}}|0\rangle + |2\rangle$ (blue dash-dot line) for different values of $|J^Z/J^X|$ and the first 50 discrete time steps. The values of average Fidelity measure $\hat{F}$ for routing state $|0\rangle$ for there ration (red line $J^Z/J^X = 1$, green dotted $J^Z/J^X = 2$, blue dash-dot line $J^Z/J^X = 3$) are presented in plot (B) (Color figure online)

Figure 3 contains the values of Fidelity measure for three exemplary states. The time values are scaled to the time where the time variable is changed discretely each $\pi/2J^X$. It is possible to reach the Fidelity value ≈ 0.99 but it requires to select the parameters for each coupling.

5 Conclusions

The construction of quantum router was presented in this article. The router is a generalization of solutions working on qubits, and discussed in the literature in therms of spin interactions between quantum units of information. As the set of operators, we utilize the Gell-Mann matrices which are qutrit generalization of the Pauli operators. It is necessary to emphasize that the used Hamiltonian allows indicating the ways of possible physical implementation. The Hamiltonian describes interactions given by the generalized Pauli group, i.e. Gell-Mann operators for qutrits, to the physical realization of the router.

We have briefly shown that joining the routers allows building the structures able to transfer a quantum state to the defined node in a quantum network.

It is possible to achieve very high accuracy of information transfer in the router. However, it requires to carefully select the coupling parameters. The obtained values of the average Fidelity measure (≈ 0.99), show that the router operates correctly.

An interesting direction for further work is a hybrid system which could transfer the qubit state to one specific output from the available outputs with a qudit controlling state.

Acknowledgments. We would like to thank for useful discussions with the *Q-INFO* group at the Institute of Control and Computation Engineering (ISSI) of the University of Zielona Góra, Poland. We would like also to thank to anonymous referees for useful comments on the preliminary version of this chapter. The numerical results were done using the hardware and software available at the "GPU μ-Lab" located at the Institute of Control and Computation Engineering of the University of Zielona Góra, Poland.

References

1. Behera, B.K., Seth, S., Das, A., Panigrahi, P.K.: Demonstration of entanglement purification and swapping protocol to design quantum repeater in IBM quantum computer. arXiv:1712.00854 [quant-ph] (2017)
2. Behera, B.K., Reza, T., Gupta, A., Panigrahi, P.K.: Designing quantum router in IBM quantum computer. Quantum Inf. Process. **18**(11), 1–13 (2019). https://doi.org/10.1007/s11128-019-2436-x
3. Bergmann, M., van Loock, P.: Hybrid quantum repeater for qudits. Phys. Rev. A **99**, 032349 (2019)
4. Briegel, H.J., Dür, W., Cirac, J.I., Zoller, P.: Quantum repeaters: the role of imperfect local operations in quantum communication. Phys. Rev. Lett. **81**, 5932 (1998)
5. Bruderer, M., Franke, K., Ragg, S., Belzig, W., Obreschkow, D.: Exploiting boundary states of imperfect spin chains for high-fidelity state transfer. Phys. Rev. A **85**, 022312 (2012)

6. Caleffi, M.: Optimal routing for quantum networks. IEEE Access **5**, 22299–22312 (2017)
7. Christensen, K.S., Rasmussen, S.E., Petrosyan, D., Zinner, N.T.: Coherent router for quantum networks with superconducting qubits. Phys. Rev. Res. **2**, 013004 (2020)
8. Dahlberg, A., Wehner, S.: SimulaQron—a simulator for developing quantum Internet software. Quantum Sci. Technol. **4**(1), 015001 (2018). https://doi.org/10.1088/2058-9565/aad56e
9. Diadamo, S., Nözel, J., Zanger, B., Bese, M.M.: QuNetSim: a software framework for quantum networks. arXiv:abs/2003.06397 (2020)
10. Diadamo, S., Nözel, J., Zanger, B., Bese, M.M.: Github repository (2020). https://github.com/tqsd/QuNetSim
11. Domino, K., Gawron, P.: An algorithm for arbitrary-order cumulant tensor calculation in a sliding window of data streams. Int. J. Appl. Math. Comput. Sci. **29**(1), 195–206 (2019)
12. Einstein, A., Podolsky, B., Rosen, N.: Can quantum-mechanical description of physical reality be considered complete? Phys. Rev. **47**, 777–780 (1935)
13. Erhard, M., Malik, M., Zeilinger, A.: A quantum router for high-dimensional entanglement. Quantum Sci. Technol. **2**(1), 014001 (2017). https://doi.org/10.1088/2058-9565/aa5917
14. Goswami, K., et al.: Indefinite causal order in a quantum switch. Phys. Rev. Lett. **121**, 090503 (2018)
15. Gyongyosi, L., Imre, S.: Entanglement-gradient routing for quantum networks. Sci. Rep. **7**, 14255 (2017)
16. Hahn, F., Pappa, A., Eisert, J.: Quantum network routing and local complementation. npj Quantum Inf. **5**, 76 (2019)
17. Hall, B.C.: Lie Groups, Lie Algebras, and Representations: An Elementary Introduction. Springer, New York (2003). https://doi.org/10.1007/978-3-319-13467-3
18. Hu, C.: Photonic transistor and router using a single quantum-dot-confined spin in a single-sided optical microcavity. Sci. Rep. **7**, 45582 (2017)
19. Jankowski, N., Linowiecki, R.: A fast neural network learning algorithm with approximate singular value decomposition. Int. J. Appl. Math. Comput. Sci. **29**(3), 581–594 (2019)
20. Luo, Y.H., et al.: Quantum teleportation in high dimensions. Phys. Rev. Lett. **123**, 070505 (2019)
21. Marchukov, O., Volosniev, A., Valiente, M., et al.: Quantum spin transistor with a Heisenberg spin chain. Nat. Commun. **7**, 13070 (2016)
22. Nielsen, M.A., Chuang, I.L.: Quantum Computation and Quantum Information. Cambridge University Press, Cambridge (2000)
23. Pant, M., Krovi, H., Towsley, D., et al.: Routing entanglement in the quantum Internet. npj Quantum Inf. **5**, 25 (2019)
24. Pedersen, L.H., Møller, N.M., Mølmer, K.: Fidelity of quantum operations. Phys. Lett. 384 A **367**, 47–51 (2007)
25. Plenio, M.B., Hartley, J., Eisert, J.: Dynamics and manipulation of entanglement in coupled harmonic systems with many degrees of freedom. New J. Phys. **6**(1), 36 (2004)
26. Rasmussen, S.E., Christensen, K.S., Zinner, N.T.: Controllable two-qubit swapping gate using superconducting circuits. Phys. Rev. B **99**, 134508 (2019)
27. Van Meter, R.: Quantum Networking. Wiley, Hoboken (2014). https://doi.org/10.1002/9781118648919

28. Wallnöfer, J., Zwerger, M., Muschik, C., Sangouard, N., Dür, W.: Two-dimensional quantum repeaters. Phys. Rev. A **94**, 052307 (2016)
29. Wang, M., Chen, X., Luo, S., et al.: Efficient entanglement channel construction schemes for a theoretical quantum network model with d-level system. Quantum Inf. Process. **11**, 1715–1739 (2012)
30. Zhukov, A.A., Kiktenko, E.O., Elistratov, A.A., Pogosov, W.V., Lozovik, Y.E.: Quantum communication protocols as a benchmark for programmable quantum computers. Quantum Inf. Process. **18**(1), 1–23 (2018). https://doi.org/10.1007/s11128-018-2144-y
31. Zwerger, M., Lanyon, B.P., Northup, T.E., Muschik, C.A., Dür, W., Sangouard, N.: Quantum repeaters based on trapped ions with decoherence-free subspace encoding. Quantum Sci. Technol. **2**(4), 044001 (2017). https://doi.org/10.1088/2058-9565/aa7983

Development of High Performance Computing Systems

Stanisław Kozielski[✉] and Dariusz Mrozek

Department of Applied Informatics, Faculty of Automatic Control,
Electronics and Computer Science, Silesian University of Technology,
Akademicka 16, 44-100 Gliwice, Poland
{stanislaw.kozielski,dariusz.mrozek}@polsl.pl

Abstract. The aim of the work is to present the development trends of
high performance computers. The analysis focused on system architecture, processors and computing accelerators used. Particular attention
was paid to interconnection networks, connecting system nodes. The
problem of energy saving in the construction of large systems was also
discussed. The data from the list of 500 largest computer systems was
used in the analysis.

Keywords: High performance computer · Computing accelerator ·
Interconnection network

1 Introduction

The paper discusses the observed trends in the development of high performance
computing (HPC) systems. The discussion noted both the constantly observed
progress in technologies and the less frequently occurring changes in solution
concepts. The presented analysis includes the following features of the evaluated
systems: system architecture, processors used in the construction of these systems, computing accelerators appearing more and more often in such solutions
and networks connecting nodes of computer systems. A regular source of data
on such computers is a list of 500 largest computer systems (TOP500 Supercomputer Sites) [16] compiled twice a year (since 1993).

The issue of the amount of electricity that these systems consume is increasingly important in large systems. Comparative data on this topic was collected on
the basis of The Green500 list, the list of computers with the highest energy efficiency [15]. The problems of saving energy consumed by supercomputers occupy
an increasingly important place in the design of new systems [10].

Both of the above-mentioned lists are a source of data for publications on
the development of high performance computing systems. As an example, the
article [8] can be mentioned, which focuses on the differences in the performance
of homogeneous (not using computational accelerators) and heterogeneous (using
accelerators) supercomputers. Additionally, the processors and interconnection
networks used in these supercomputers are analysed in this article.

© Springer Nature Switzerland AG 2020
P. Gaj et al. (Eds.): CN 2020, CCIS 1231, pp. 52–63, 2020.
https://doi.org/10.1007/978-3-030-50719-0_5

The development of the largest supercomputers in recent years has been closely followed in the perspective of building exascale systems [18], which will create significantly new opportunities for computational sciences.

The following sections of this paper present the major features of large computer systems and discuss their development observed in recent years.

2   Architecture of High Performance Computer Systems

The general structure of high performance computers classifies them in the category of distributed memory multiprocessor systems. The basis of the HPC systems architecture are nodes consisting of several (usually 2 or 4) processors (currently multicore) with their local memory. Such a node is then a shared-memory multiprocessor. A high performance computer therefore creates a system with a hierarchical, heterogeneous structure. It is based on nodes (shared-memory multiprocessors) connected by a interconnection network enabling processes implemented in nodes to exchange messages. The development of such systems has led to the formation of two variants.

The first solution is characterized by a very large number of nodes connected by a interconnection network with unique properties. Typically, each manufacturer has its own interconnection network solution. Such systems are called Massively Parallel Processor (MPP) [14].

In the second, more economical option, "off the shelf" servers are used as nodes, and interconnection network standards are also used. It should be added that the servers are shared-memory multiprocessors (SMP - Symmetric Multiprocessor). In this variant usually one, two or four processor servers are used. Such systems are called computer clusters [9].

It should be emphasized that the scalability of both variants is similar - limited only by the size of the interconnection networks. Clusters and MPP systems appear equally often at the top of the Top500 list. It can be added that distinguishing the name "MPP system" does not always take place and rarely occurs on the commercial market. However, this category is used in the classification of systems listed on the Top500 list [7].

In summary, the general organization of MPP systems and computer clusters is similar, the differences relate primarily to technological solutions. A more detailed discussion of these differences will take place in the following sections.

As noted above, clusters are generally a cheaper solution than MPP systems with similar performance. Therefore, the tendency to standardization observed in the construction of computer systems leads to a gradual decrease in the number of MPP systems on the Top500 list. This is illustrated in Table 1. Data presented in this table (and in other tables) were collected from the Top500 lists available in November of a given year.

It can be seen that despite the decreasing number of MPP systems on the Top500 list, their share in the top ten systems of this list is significant.

It should also be emphasized that the average MPP system has more performance than the average cluster. In the recent ranking (November 2019), the

Table 1. Number of MPP systems and clusters in subsequent editions of the Top500 list

Year	MPP systems	Top MPP system positions	Clusters	Top cluster positions
2000	346	1–30	143	31, 32, 44
2005	103	1, 2, 3, 4, 6	397	5, 8, 11, 15
2010	85	1, 2, 5, 8, 9	415	3, 4, 6, 7, 14
2015	74	2, 3, 5, 6, 7	426	1, 4, 10, 13,15
2019	42	3, 6, 7, 12, 13	458	1, 2, 4, 5, 8

number of MPP systems was only 8.4% of the number of all Top500 list systems, while their total performance was 17.2% of the total performance of all systems.

3 Processors Used in HPC System Nodes

Intel processors dominate the largest computer systems. This is shown in Table 2.

Table 2. Processors used to build systems from the Top500 list

Year	Intel	IBM	AMD	Others
2010	390	40	58	12
2015	445	26	21	8
2016	462	22	7	9
2017	471	14	5	10
2018	476	8	3	13
2019	474	14	5	7

Table 2 shows the dominance of Intel processors among the Top500 systems, especially in recent years, exceeding 90% of the number of systems. On the recent ranking (November 2019), the systems with Intel Xeon Scalable (Gold - 165, Platinum - 40) and Xeon E5 (Broadwell - 183) processors are most frequently represented.

Intel Xeon processors are used in both clusters and MPP systems. Cluster nodes usually contain 2, less often 4 processors (e.g. Lenovo SR650, Sugon TC6600, Inspur NF5468M5 clusters). Similarly, the node of the sample MPP system, the Cray XC40 computer (built by Cray/HPE), contains 2 Intel Xeon processors.

The increase in the share of IBM processors in the recent ranking of the Top500 list is due to the introduction of several new systems based on the Power9 processor.

The IBM Power9 processor is the basis of the two largest systems on the Top500 list: Summit and Sierra. The node in the Summit system is the IBM Power System AC922 server, containing 2 Power9 processors, while the node of the Sierra system is the IBM Power System S922LC server, also containing 2 Power9 processors (these nodes also contain accelerators, which we will discuss in the next section).

The cases discussed relate to processors widely used in the production of computers of various classes. The attention can be also paid to the unique processors used only in MPP supercomputers. Sunway SW26010 is the chosen example of such a processor (included in Table 2 in the "Others" column). Sunway SW26010 processor was used in the Sunway TaihuLight system, which took first place on the Top500 ranking in 2016–2018. This 64-bit RISC processor consists of 4 core sets, including 1 management core and 64 computing cores. Therefore, the entire processor has 260 64-bit cores. These cores perform, among others, vector instructions (according to SIMD model) on 256-bit data.

In the summary of this point, it should be emphasized that new models of supercomputers usually use new models of processors with the largest number of cores. The higher density of core packing makes it possible to reduce the size of systems' structures, facilitates solving problems of cooling computer modules, and also facilitates the construction of interconnection networks, connecting system nodes.

4   Computing Accelerators

An increasing number of supercomputers are equipped with computing accelerators. NVIDIA's accelerators, which are the result of the company's introduction of the CUDA (Compute Unified Device Architecture) architecture into its graphics processors, appeared on the Top500 list in 2010 and since then their number on this list, with slight fluctuations, has been constantly increasing.

The calculations carried out in this architecture according to the SIMD model ensure obtaining very high performance, but only in tasks characterized by the data level parallelism, e.g., vector and matrix calculations. Such calculations are also typical for computer graphics, hence the symbiosis of the CUDA model with graphics processors.

The SIMD model is also the basis for vector instructions that extend the classic instruction lists. Such processor architecture development is represented, among others, by Intel Xeon Phi coprocessor, which is one of the alternative forms of accelerators. In addition, it represents among Intel processors so-called MIC architecture (Many Integrated Core), characterized by a large number of cores (60–70). On some HPC systems, this processor is used as a stand-alone main processor. In addition, combinations of NVIDIA and Intel Xeon Phi accelerators are also used in supercomputers.

An interesting concept of accelerator was implemented by PEZY Computing company. The basis of the PEZY processor (accelerator) is the core (PE - Processing Element), which is a superscalar processor, implementing 8-track SMT multi-threading. The latest version of the PEZY-SC2 accelerator contains 2048 cores, managed by 6 P-Class P6600 MIPS processors (MIPS64R6 architecture).

The use of computing accelerators in computer systems, which in recent years were on the Top500 list, is presented in Table 3. The number of systems containing a combination of NVIDIA and Intel Xeon Phi accelerators were distinguished as a component of the sum in both columns *NVIDIA - totally* and *Intel Xeon Phi.*

Table 3. Computing accelerators in supercomputers

Year	Number of systems with accelerators	NVIDIA			Intel Xeon Phi	PEZY	Others
		Totally	Including				
			P100	V100			
2010	17	9					8
2015	104	66 + 4			28 + 4		6
2016	86	60 + 3	2		21 + 3	1	1
2017	105	86 + 2	47	1	12 + 2	5	
2018	142	118 + 3	64	46	17 + 3	2	2
2019	154	134 + 2	32	96	14 + 2	1	3

The data in Table 3 show a steady increase, with slight deviations, in the number of systems using computing accelerators. On the other hand, this number only reaches 30% of the entire list, which may be a bit surprising. Accelerators are definitely dominated by NVIDIA products, the most advanced Pascal and Volta solutions in the last two years. While the share of NVIDIA accelerators tends to increase significantly, the share of Intel Xeon Phi is rather decreasing.

It is interesting to analyse the use of accelerators in the largest systems, placed at the top of the Top500 list. There are 5 systems using NVIDIA accelerators and 1 system with Intel Xeon Phi in the top ten systems of the ranking. Within the second ten: 2 - NVIDIA, 5 - Intel Xeon Phi can be found, whereas, the calculations for the third ten repeat the results for the first one.

It is interesting to analyse the share of accelerators in the total performance of the two largest systems on the Top500 list: Summit and Sierra, partly already presented in the previous section of this paper. The node of the Summit system is the IBM Power System AC922 server, containing 2 Power9 processors and 6 NVIDIA Tesla V100 accelerators. The Sierra system node is the IBM Power System S922LC server, also containing 2 Power9 processors and 4 NVIDIA Tesla V100 accelerators.

Data presented by Lavrence Livermore National Laboratory, where the Sierra supercomputer [11] is located, allow to assess the share of accelerator performance in the total performance of this computer:

- number of nodes = 4 320,
- number of CPU (Power9) = 8 640,
- number of GPUs (Tesla V100) = 17280,
- GPU performance = 7 TFlop/s.

The total theoretical performance (peak performance) is:

- for all CPUs: Rpeak CPU = 4 666 TFlop/s
- for all GPUs: Rpeak GPU = 120 960 TFlop/s
- for Sierra supercomputer: Rpeak = 125 626 TFlop/s

Thus, the share of accelerators in the theoretical (peak) performance of the Sierra supercomputer is 96.28%!

For the largest Summit supercomputer, similar data is [12,17]:

- total theoretical performance of GPU = 193 536 TFlop/s
- theoretical performance of Summit supercomputer = 200 794.9 TFlop/s

The share of accelerators in the theoretical (peak) performance of the Summit supercomputer is 96.38%!

In the summary, we will additionally note, that the recent development of the architecture of computing accelerators includes, among others the needs of intensively developed artificial intelligence. Introduced to new models of accelerators, the so-called tensor cores, performing operations of matrix multiplication and addition, support some computer graphics operations (ray tracing), but also facilitate the implementation of deep learning algorithms and simulation of neural networks.

5   Interconnection Networks in HPC Systems

In computer systems, communication problems affect all elements of these systems at many levels. Detailed considerations can therefore relate to communication within the chip (processor), communication within the node (server), as well as communication between nodes. As explained in point 2 of this work, supercomputer nodes are shared-memory multiprocessors and their construction and internal communication are not the subject of this work. At this point, we will consider the specifics of interconnection networks, connecting nodes of supercomputer systems. As presented in point 2, the supercomputers considered represent the architecture of clusters (mostly) or MPP systems.

On the cluster market, in the last several years two interconnection network standards have competed with each other: Ethernet and InfiniBand, and some years ago they were also joined by the Omni-Path standard. These interconnection networks compete with each other with two technical parameters: bandwidth and switching latency, and of course the price. When it appeared on the market, InfiniBand clearly outperformed its Ethernet infrastructure, and these solutions were clearly more expensive. In the following years, the development of Gigabit Ethernet and 10 Gigabit Ethernet standards closed these differences, but InfiniBand continues to provide better performance. A few years ago, a new standard called Omni-Path was launched by Intel [4]. The development of interconnection networks is reflected in Table 4, which shows the number of Top500 systems using the main standards of interconnection networks in recent years.

Table 4. Interconnection network standards in Top500 systems

Year	Gigabit Ethernet		InfiniBand		Omni-Path		Custom Interconnect		Proprietary network	
	Total	Top	Total	Top	Total	Top	Total	Top	Total	Top
2000	16	80, 126, 211								
2005	250	65, 89, 95	27	5, 20, 50			7	6, 10, 14	20	1, 2, 9
2010	226	39, 115, 125	214	3, 4, 6			32	5, 10, 18	18	1, 2, 8
2015	181	65, 66, 75	237	10, 13, 15			74	1, 2, 3	8	21, 25, 30
2016	207	88, 93, 94	187	13, 16, 17	28	6, 12, 41	71	1, 2, 3	7	30, 35, 43
2017	228	73, 101, 103	163	4, 17, 21	35	9, 12, 13	67	1, 2, 3	7	38, 43, 52
2018	252	38, 86, 96	135	1, 2, 7	43	8, 13, 14	64	3, 4, 5	6	56, 63, 74
2019	259	48, 75, 81	140	1, 2, 5	50	9, 14, 15	45	3, 4, 6	5	72, 84, 101

The content of Table 4 requires a few explanations and broader comments.

- Column *Total* contains a number of systems on the Top500 Supercomputers list.
- Column *Top* contains numbers of the first positions on the Top500 list of systems with a given network category.
- For the first three items (2000–2010) the table does not include, due to lack of space, all standards and technologies of interconnection networks used at that time. They were, among others networks: Myrinet and Quadrics - multilevel networks preceding the InfiniBand standard; SP Switch - a multi-stage network used in MPP systems by IBM; NUMAlink - a network used in SGI MPP systems.
- For the above reasons, the sum of the number of systems does not reach 500 for the first three items (2000–2010).
- In the Top500 list database, interconnection networks have been classified in recent years in five main categories included in Table 4. The data in this table for 2015–2019 correspond to these categories.

When comparing the use of Ethernet and InfiniBand standards, the following facts should be highlighted:

- In 2000, systems using Ethernet or Fast Ethernet technology were present on the Top500 list. For the remaining years, the "Gigabit Ethernet" column in Table 4 shows the number of systems with Gigabit Ethernet technology or (in later years) 10 Gigabit Ethernet.

- The Gigabit Ethernet standard (including 10 GbE) has been used since 2005 (with one exception) as the most common connecting network in systems listed on the Top500 list. However, these systems were not at the top of the list.
- The InfiniBand standard was Ethernet's biggest competitor taking the number of systems where they were applied into consideration (although it was a winner only once). In contrast, supercomputers with Infiniband have always been widely represented at the top of the Top500 list, which confirms the advantage of InfiniBand parameters.
- In the past two years, InfiniBand has been at the top of the Top500 list because it has been used in the IBM systems discussed above: Summit and Sierra. They use InfiniBand switches with EDR technology, providing a throughput of 100 Gb/s.

Cluster systems use almost exclusively Gigabit Ethernet, InfiniBand or Omni-Path standards in recent years. Omni-Path network parameters are between Ethernet and InfiniBand parameters. The most popular topology for these networks is the so-called Fat tree [13], in which network switches are combined into a tree, and the throughput of subsequent tree levels increases as the root approaches.

Clusters not using the above standards are exceptions. One of the few are Fujitsu company clusters using the Tofu Interconnect 2 network. The topology of this network is determined by a six-dimensional torus [1]. This dimension of the network was used, among others in order to obtain greater maximum throughput than in the three-dimensional torus, used e.g. in Cray networks, discussed below. Among the distinguishing features of the Tofu 2 network can also be mentioned the integration of this network switch with the processor used (SPARC64) in one chip. The Tofu network is still being developed [2]. The Tofu network has been included on the Top500 list within the "Proprietary Network" category.

The last category of interconnection networks in Table 4 that has not yet been discussed is "Custom Interconnect". Closer analysis of the systems in this category shows that in recent years, except for a few clusters, it covers almost all MPP systems. Due to the specificity of the networks in these systems, we will present selected solutions.

Among the MPP systems, Cray company systems are the most numerous. The distinctive feature of these systems are the original interconnection networks. In earlier systems, the company used its own switches (SeaStar and Gemini) with very high bandwidth, connected by a network with a three-dimensional torus topology. A new solution introduced by this company is Aries switch technology and change of the interconnection network topology to Dragonfly [3].

The Dragonfly topology has a two-level hierarchical structure. At the first level, the nodes are divided into small groups, the nodes in the group are connected according to one of the classic topologies, e.g. Fat tree, torus, Butterfly network and others. The second level creates a interconnection network connecting groups. In this network, the adopted solution is to combine each group with all others.

In comparison with the Fat tree, the Dragonfly network enables, among others reducing the number of optical links (connecting groups) and reducing the number of hops in the message path when making connections. Additionally, it facilitates the implementation of adaptive routing.

Aries switches with Dragonfly topology are used in the intensively developed XC series of Cray supercomputers (XC30, XC40, XC50). 34 such systems are present on the last Top500 list.

Among the other interconnection network solutions in the MPP systems present on the Top500 list the network of the IBM BluGene/Q supercomputer can be mentioned. The topology of this network is a five-dimensional torus. The choice of such topology and this dimension was justified by the following arguments [5]: (1) achieving high throughput of each node with its nearest neighbors and reducing the maximum hop count, compared to a lower dimension torus; (2) the ability to divide the machine into partitions in which the applications minimally affect each other; (3) the ability to minimize optical connections and implementation of the most of connections as electrical ones, which reduced costs. In recent years, however, there is a lack of data indicating the further development of this line of supercomputers.

6   Energy-Efficient Supercomputers

Energy consumption has become one of the important problems in the further development of high-performance computer systems. The Summit supercomputer, currently with the highest performance, consumes 10.09 MW of electricity, but its predecessors were even more energy intensive, as shown in Table 5.

Table 5. Selected supercomputers with the highest energy consumption

Name	Vendor	Electrical power [MW]	Performance [PFlop/s]	Top500 position
Tianhe-2A	NUDT, China	18.482	61.444	3rd - 2018
		17,808	33.862	1st - 2015
Sunway TaihuLight	NRCPC, China	15.371	93.014	1st - 2016
K computer	Fujitsu, Japan	12.659	10.51	1st - 2011
Summit	IBM, USA	10.096	148.6	1st - 2019

Development of energy-saving computers became of interest of computer vendors in recent years and the progress of this work for supercomputer systems is tracked on the Green500 list. It is a ranking of computers from the Top500 list, according to the ratio (GFlop/s)/W, that is the quotient of the computer's performance and its electrical power consumption. The Green500 list has been published since 2007, twice a year, similar to Top500. Selected data describing the leaders of the Green500 lists is presented in Table 6.

Table 6. Selected leading systems on the Green500 lists

Year	System	Performance [TFlop/s]	Electrical power [kW]	Power efficiency [GFlops/W]	Top500 position
2007	Blue Gene/P, IBM, Great Britain	11.1	31	0.358	122
2010	Blue Gene/Q Prototype, IBM, USA	65.3	38.8	1.684	116
2015	ExaScaler-1.4 PEZY-SC, Japan	605.6	86.12	7.0314	135
2016	NVIDIA DGX-1, NVIDIA Tesla P100, USA	3307	349.5	9.462	28
2017	Shoubu system B - ZettaScaler-2.2, PEZY Computing/Exascaler Inc., Japan	842	50	17.009	259
2018	Shoubu system B - ZettaScaler-2.2, PEZY Computing/Exascaler Inc., Japan	1 063.3	60	17.604	375
2019	A64FX prototype, Fujitsu, Japan	1 999.5	118	16.876	159

Assessing the data presented in Table 6 it can be noticed that in the years 2007–2019, that is during the period recorded on the Green500 list, the energy efficiency of the most economical computers increased 49 times, while the performance of the fastest computers increased 310 times. At the same time, the analysed data show that the most economical computers occupied quite distant positions on the Top500 list, thus, their performance is not yet competitive with the systems from the top of this list. At this point, it should be noted that at the 5th place of the Green500 list at the end of 2019 is currently the most powerful supercomputer - Summit. Thus, energy efficiency is also becoming a feature of the largest computer systems.

The most energy-efficient system in Table 6 called Shoubu - ZettaScaler contains two key solutions: original PEZY processors built by PEZY Computing and a cooling system constructed by Exascaler Inc. The nodes of this system include the Intel Xeon D processor (16 cores) and 8 PEZY-SC2 processors (presented in point 4 of this paper), that act as accelerators.

The technical construction of the whole system includes many tanks, filled with an electrically neutral fluorocarbon-based liquid, in which densely packed computer modules are immersed. The flowing liquid cools the computer nodes intensively.

The Fujitsu A64FX computer, which is the leader of the last Green500 list, is based on the A64FX processor developed by Fujitsu. This 48-core processor represents the ARM architecture and performs vector instructions according to the 512 bit SIMD model. The Fujitsu A64FX system also uses the Tofu interconnection network discussed above.

7 Summary

The paper presents an analysis of trends observed in recent years in the development of supercomputers on the basis of data available on the TOP500 Supercomputer Sites lists. Significant impact on the development of supercomputers have: continuous development of processor technology, in particular increasing the number of cores in processors; dynamic development of computing accelerators; development of new topologies and technology of interconnection networks connecting system nodes. The introduction of energy-efficient computer designs is becoming a very important factor in the development of supercomputers.

The main goal of supercomputer development in recent years has become the construction of exascale systems [6]. Building a system with 1 EFlop/s performance (i.e. 10^{18} Flop/s) will be a big event in the world of supercomputers.

Acknowledgments. The work was carried out within the statutory research project of the Department of Applied Informatics, Faculty of Automatic Control, Electronics and Computer Science, Silesian University of Technology.

References

1. Ajima, Y., et al.: The Tofu interconnect D. In: 2018 IEEE International Conference on Cluster Computing (CLUSTER), pp. 646–654, September 2018. https://doi.org/10.1109/CLUSTER.2018.00090
2. Ajima, Y., et al.: Tofu interconnect 2: system-on-chip integration of high-performance interconnect. In: Kunkel, J.M., Ludwig, T., Meuer, H.W. (eds.) ISC 2014. LNCS, vol. 8488, pp. 498–507. Springer, Cham (2014). https://doi.org/10.1007/978-3-319-07518-1_35
3. Alverson, B., Froese, E., Kaplan, L., Roweth, D.: Cray XC series network. In: White Paper WP-Aries01-1112, Cray Inc. (2012)
4. Birrittella, M.S., et al.: Intel® omni-path architecture: enabling scalable, high performance fabrics. In: 2015 IEEE 23rd Annual Symposium on High-Performance Interconnects, pp. 1–9, August 2015. https://doi.org/10.1109/HOTI.2015.22
5. Chen, D., et al.: The IBM Blue Gene/Q interconnection fabric. IEEE Micro **32**(1), 32–43 (2012). https://doi.org/10.1109/MM.2011.96
6. Conway, S., Joseph, E., Sorensen, B., Norton, A.: Exascale innovations will benefit all HPC users. In: Hyperion Research, White Paper HR4.0011.03.14.2018 (2018)
7. Dongarra, J., Sterling, T., Simon, H., Strohmaier, E.: High-performance computing: clusters, constellations, MPPs, and future directions. Comput. Sci. Eng. **7**(2), 51–59 (2005). https://doi.org/10.1109/MCSE.2005.34

8. Gao, Y., Zhang, P.: A survey of homogeneous and heterogeneous system architectures in high performance computing. In: 2016 IEEE International Conference on Smart Cloud (SmartCloud), pp. 170–175, November 2016. https://doi.org/10.1109/SmartCloud.2016.36

9. Hennessy, J.L., Patterson, D.A.: Computer Architecture: A Quantitative Approach. Morgan Kauffman, Waltham (2012)

10. Hood, R.T., Mehrotra, P., Thigpen, W.W., Tanner, C.B., Buchanan, C.J., Chan, D.S.: Expanding a supercomputer facility using modular data center technology, Conference Paper: SC17, 12–17 November 2018, Denver, CO, United States. NASA Technical Report Server, Report ARC-E-DAA-TN41643 (2018). https://ntrs.nasa.gov/search.jsp?R=20180007549. Accessed Jan 2020

11. L.L.N. Laboratory: Sierra (2020). https://hpc.llnl.gov/hardware/platforms/sierra. Accessed Jan 2020

12. O.R.N. Laboratory: Summit (2020). https://www.olcf.ornl.gov/olcf-resources/compute-systems/summit/. Accessed Jan 2020

13. Leiserson, C.E.: Fat-trees: universal networks for hardware-efficient supercomputing. IEEE Trans. Comput. **C−34**(10), 892–901 (1985). https://doi.org/10.1109/TC.1985.6312192

14. Tanenbaum, A.S.: Structured Computer Organization. Pearson Education India, Delhi (2016)

15. The TOP500 project: The green500 list (2020). https://www.top500.org/green500/. Accessed Nov 2019

16. The TOP500 project: The TOP500 list (2020). https://www.top500.org/. Accessed Jan 2020

17. Top500: Summit (2019). https://www.top500.org/system/179397. Accessed Jan 2020

18. Wright, S.A.: Performance modeling, benchmarking and simulation of highperformance computing systems. Fut. Gener. Comput. Syst. **92**, 900–902 (2019). https://doi.org/10.1016/j.future.2018.11.020. http://www.sciencedirect.com/science/article/pii/S0167739X18328590

Peer-to-Peer Transfers for Crowd Monitoring - A Reality Check

Christin Groba[✉] and Alexander Schill[✉]

Chair of Computer Networks, Technische Universität Dresden, Dresden, Germany
{Christin.Groba,Alexander.Schill}@tu-dresden.de

Abstract. Peer-to-peer transfers allow for sharing crowd monitoring data despite the loss of network connectivity. However, limited insight into real-world deployment contexts can let the protocol design go astray - particularly, if a certain nature of participant behaviour and connectivity changes is assumed. This paper focuses on the delivery of crowd monitoring data. It puts a protocol out for a reality check that switches to peer-to-peer (p2p) communication when the infrastructure network connection is lost. The evaluation at an annual indoor fair asked visitors to make their phones visible to peers, run the protocol, and share crowd monitoring data. The results show that most of the participants formed a large radio cluster throughout the event. This made p2p networking only possible and enabled a more robust upload of crowd monitoring data. However, dynamic switching between infrastructure network and p2p communication also increased the volatility of the system, calling for future optimizations. The presented measurement results provide further insights into these details.

Keywords: Crowd monitoring · Peer-to-peer · Bluetooth · Android · Real-world evaluation

1 Introduction

Awareness of how event visitors roam the venue allows for aligning safety measures and for optimizing the event setup. Monitoring techniques that engage people in the crowd to share sensor data from their phones is a novel way to infer crowd metrics without expensive camera deployments or crowd stewards.

However, due to the crowd density, the demand for networking bandwidth increases while the infrastructure is designed for the average case that does not accommodate for crowds of people. Crowd monitoring data then competes for bandwidth just like any other traffic and participants of the crowd monitoring campaign may not be able to share their data with the campaign server. Off-loading to peers that still have a connection allows for sharing such data nonetheless. Our previous work shows that going as far as using peer-to-peer

This research is funded by the German Research Foundation (DFG) under GR 4517/1-1.

Fig. 1. An annual indoor fair is the setting for testing the protocol.

(p2p) data forwarding on default incurs long data delays and therefore infrastructure network connectivity should be used if available [4].

This paper investigates a protocol that switches to p2p data transfer only when a device lost network connectivity. The device then attempts to offload its data to a one-hop peer that is still connected to the infrastructure. During protocol design, however, one makes assumptions (at times unconsciously) about the way participants behave or the network connectivity changes. This raises the question of how the protocol performs in a real-world deployment where participation, user behaviour, and network infrastructure are beyond the designer's control. The evaluation of the protocol is set during an annual indoor fair in one of the buildings on university campus (cp. Fig. 1). Over the course of six and a half hours 47 visitors participated and shared location, connectivity and peer-related data.

The contribution of this paper is an in-depth analysis of the challenges a real-world crowd monitoring scenario has on the design of a p2p data transfer protocol. The paper explores the impact of participation patterns, clusters of visitors, and implications of unexpected network behaviour.

2   Related Work

Participatory crowd monitoring relies on people in the crowd to share crowd monitoring data with a campaign server. GPS readings, acceleration data, or Wifi fingerprints allow for inferring the crowd's density and flow [9] as well as groups of people that move as a cohesive whole [6]. Experience from large-scale events show that network access via the existing infrastructure becomes slow and at times even impossible [2]. The resulting long delays for crowd monitoring data call for offloading solutions to reduce the load on the network [1]. In our case, participants experienced a high rate of connectivity changes with networking interfaces frequently switching between connected and disconnected. The paper shows, how such an unexpected network behaviour affected the tested protocol.

One way to reduce network load is to establish the crowd metric among all peers and assign one peer that uploads the metric to the server. UrbanCount [3], for example, proposes a distributed crowd counting technique. It builds on an epidemic model where nodes receive radio signals from other nodes and broadcast a list of "seen" nodes. Trace-driven simulations show that such an approach produces a precise count when the crowd is dense. A similar approach [5] based on audio tones shows high scalability and accuracy at much less energy consumption compared to radio-based solutions. As for peer assignment strategies, techniques from collaborative sensing may be adapted that lets mobile nodes decide for themselves whether to become a peer with additional responsibilities [8]. A stochastic algorithm makes this decision in intervals and ensures a fair and effective allocation.

This paper takes another approach by offloading data to a peer only when connectivity issues occur. However, instead of advancing to complex solutions as proposed for domains other than crowd monitoring [11], it first focuses on early evaluation in a real-world deployment.

3   P2P Protocol

The protocol is based on the assumption that a set of nodes representing people with modern phones run an app to participate in a crowd monitoring campaign. They arbitrarily move around while visiting an event, for example, a fair. Each node has a certain capacity to store a number of sensor data and to perform the following tasks:

- Discover available peers in vicinity and make itself visible to other peers,
- Accept connection requests and store data received from peers, and
- Upload data, if a connection to the crowd monitoring server is available.

Further, it is assumed that some nodes experience network disconnects due to the density of nodes. These nodes rely on transferring their data to peers that are able to upload directly to the server via their WiFi or cellular interface.

When the protocol notices a connectivity change event, it switches depending on the change to either the p2p part or the direct upload part (cp. Fig. 2). The p2p part of the protocol starts the peer discovery and stops it as soon as it finds a one-hop peer that is connected. This spares a selection phase, which, when peer availability changes in the meantime, leads to peer connection issues. The node transfers its data and removes its local copy once the transfer is completed. If the connection is closed prematurely, the transfer is aborted, and the data remains at the node. Further, if no connected peer is found, discovery times out and data

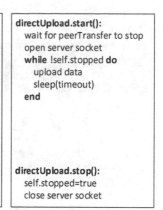

```
onConnectivityChange:
  if connection lost then
    directUpload.stop()
    peerTransfer.start()
  else
    peerTransfer.stop()
    directUpload.start()
  end
```

```
peerTransfer.start():
  wait for directUpload to stop
  while !self.stopped do
    start peer discovery
    onConnectedPeerFound:
      stop peer discovery
      transfer data to peer
    onPeerDiscoveryTimeout:
      save data
      sleep(timeout)
  end

peerTransfer.stop():
  self.stopped=true
  stop peer discovery
```

```
directUpload.start():
  wait for peerTransfer to stop
  open server socket
  while !self.stopped do
    upload data
    sleep(timeout)
  end

directUpload.stop():
  self.stopped=true
  close server socket
```

Fig. 2. Protocol pseudocode

stays stored locally. Afterwards, the node idles for a given timeout. The direct upload part of the protocol opens a server socket and in intervals uploads data collected locally and received from peers directly via the infrastructure. When this part is stopped, the server socket closes.

The protocol is in either of two states: peer transfer or direct upload. Given that a state transition is triggered solely by a connectivity change i.e., connected or disconnected, the protocol is considered deterministic. That trigger, however, is managed outside the protocol and may be seen as a shared resource. Locally the protocol is nonetheless deadlock-free, as it remains fully active in its current state until it receives a change notification. From a distributed perspective, the protocol is also deadlock-free, since none of the peers waits for another peer to take action. In case connection requests are not granted or messages do not get delivered, the underlying communication protocol indicates this failure and the node is free to abort or retry communication, possibly with another peer. Timeouts ensure that a node makes progress within a state such as terminating discovery if no adequate peer is found or stop idling to upload data again. With regard to testing, preliminary tests in the laboratory and with a small number of students ensured that the implementation of the protocol is bug-free and meets the research objectives.

The communication among phones is implemented with classic Bluetooth because it is widely available on Android phones and can be used without restricting the regular Internet access. In terms of integrating iOS devices and allowing for cross-platform communication, we also implemented the protocol with Bluetooth Low Energy. While lab tests show promising results, the evaluation in a real-world setting is still on-going. Meanwhile, this paper shares experiences with classic Bluetooth. Phones need to authorize their Bluetooth visibility to become discoverable by peers. Bluetooth discovery, however, cannot be configured to transmit protocol-specific data, namely the connectivity status of the phone. As a workaround, the phone's Bluetooth name is edited,

(a)   (b)

Fig. 3. The app accompanying the annual fair OUTPUT.DD provides an opt-in feature for participating in the mobile crowd monitoring experiment and evaluating the p2p protocol (a). Participants have exclusive access to a heat map that color-codes visitor densities in different parts of the event area (b).

which as part of the Bluetooth discovery beacon conveys this information. Further, the Bluetooth-based p2p communication runs over insecure connections to avoid manual pairing of phones. A security protocol like Transport Layer Security (TSL) is necessary to allow for privacy and data integrity despite insecure p2p communication. The implementation of such security measures is still future work.

4   Experiment

The experiment ran during an indoor fair in one of the main buildings on university campus. The exhibition area included an open space of 800 square meters and a number of show rooms spread across three floors. The companion app of that annual event included a opt-in feature for participating in the mobile crowd monitoring experiment and evaluating the p2p protocol (cp. Fig. 3a). The app links program items to an interactive floor plan and applies gamification to increase participation as well as the overall visitor engagement during the fair. Once enabled and all necessary permissions provided, participants shared sensor and log data via their phone's wireless communication interface with an application server in the Internet.

Modern smartphones provide the necessary resources to establish p2p connections, buffer and upload data. Participation, however, draws from the phone's mobile data plan, if Wifi access is unavailable, and consumes scarce battery resources. For latest phone generations like the Google Pixel 3, the battery level drops by two percent per hour running the protocol. More dated phones like the Nexus 5X require three percent per hour. In return for these participation costs, participants had exclusive access to a heat map, which turned the collected data into a map overlay color-coding visitor densities in different parts of the event area (cp. Fig. 3b).

Over the course of the event and among the visitors, about 89 people used the companion app and of those 47 participated in the experiment. The participants shared four sets of data: First, mock sensor readings that the phone created once a second as payload for the protocol. Second, sightings of Bluetooth Low Energy (BLE) beacons that were deployed on site to track the position of the participants. Third, sightings of peers in proximity, which contained the peer's device id, connectivity state, and protocol state. Fourth, the result of each protocol iteration, e.g., whether a p2p transfer was completed or aborted. All data was timestamped on creation and timestamped again when it arrived at the server, running the network time protocol for synchronized clocks. It was buffered locally and uploaded when the device was connected, or as in the case of the mock readings, when a p2p data transfer was possible. In a realistic scenario, the mock readings would be replaced by peer and static beacon sightings to allow for a cluster analysis similar to Fig. 5 and for identifying perilously dense or trapped areas.

In intervals a software controller changed the device's connectivity state to activate the p2p part of the protocol. That is, randomly after 40 to 60 s being connected, the device disconnected for 30 to 50 s and thereafter reconnected again. The controller worked only protocol-internally leaving the connectivity via the actual network interface unaffected. Only when the actual network connectivity changed, the control adjusted accordingly and dis/reconnected when the network interface did so.

4.1 Participation

With visitors coming and going, their participation in the experiment varies over time. Figure 4 depicts the participation behaviour as broken bars. A horizontal line of bars represents one device. Bars are broken when successive sensor readings are more than two minutes apart. The sensor readings, here, refer to the mock readings created by the phone. They reflect participation best because they are unaffected by the device's distance to real sensor sources and other peers. Some participants contribute right from the start, while others join-in later. Some hold out to the end, while others leave early or contribute only for a short period of time.

Participation gaps may occur because participants cancel their participation deliberately, e.g., out of curiosity to see the effect on app features. Or, participants pause participation involuntarily as they miss notifications on the renewal

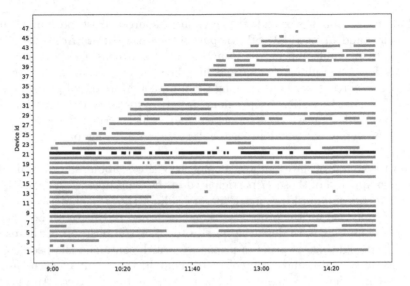

Fig. 4. Participation in the experiment varies: Some devices contribute throughout the event (e.g., device 9) while others do so only intermittently (e.g., device 21).

of app permissions. Expiring permissions, like for being Bluetooth discoverable, is among Android's efforts to protect their users' privacy. This, however, challenges apps with background features like a p2p protocol. They need to bring user attention back to the phone without negatively impacting the user's current experience in the real-world. Considering the staccato-like behaviour of the device 21 in Fig. 4, there may be other reasons, which are not yet clear.

4.2 Radio and Spatial Clusters

Over time participants formed clusters in terms of their spatial distance and radio range. Spatial clusters are derived from sightings of BLE beacons. A centroid technique [7] evaluates the beacons' received signal strength and positions each participant accordingly. Density-joining [10] clusters participants based on their neighbourhood within a two meter radius. For radio clusters, the peer sightings dataset is used, which defines neighbourhood as the number of peers a device senses in its radio range. Density-joining assigns those to the same cluster that have at least one peer in common.

Figure 5 depicts the number and size of the clusters over the course of the experiment. A dot represents a cluster. It grows in size and lightness, the more members the cluster has. In spatial terms, participants are rather scattered and independently roam the event area. This is reflected by the high number of small dots in the top part of the figure. The exception is past noon (12:20–12:50) when 10 to 15 participants gather in the same part of the event area. This may be due to a program item that catches the interest of many visitors. In contrast, the bottom of the figure shows the evolution of one large radio cluster and how it grows to more than 20 members towards the end of the event at 14:40. There are only few small radio clusters with one to five members that are disconnected from the rest possibly because they are in the most remote places of the event area. The analysis of the radio clusters shows that most participants are part of the same p2p network, which allows for data exchange and task offloading.

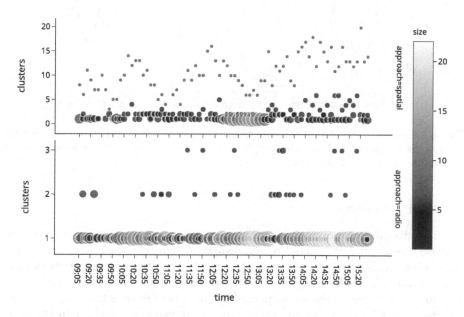

Fig. 5. Despite the participants' spatial distribution (top), most of them are part of one large radio cluster (bottom) leaving only few to small isolated radio clusters.

4.3 Connectivity Changes

With a computer science building as the venue for the event and experiment, the assumption is that network connectivity is not an issue. Disconnects that do occur, would not be sufficient to activate and sample the p2p part of the protocol in suitable quantities. A software controller thus induced artificial dis- and reconnects.

Analysing the connectivity state in the peer sightings dataset, Fig. 6 shows the median connectivity change rate over time. Considering the effect of the software controller, a rate of up to 1.5 changes per minute is expected. At times, however, the depicted values are much higher. This means, devices experience additional dis- and reconnects from the actual networking infrastructure. The peak times at 12:45 and 14:35 coincide with the times when the size of the radio and spatial clusters peak. This suggests that already at a medium size indoor event, network issues occur when the infrastructure is not particularly adjusted to the expected number of visitors. Typically, existing infrastructure is designed for the average use case that does not accommodate for crowds of people.

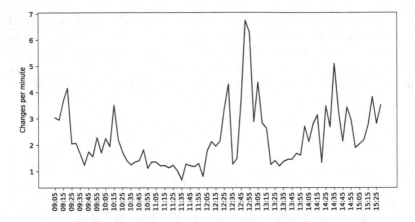

Fig. 6. The median connectivity change rate is at times much higher than the expected 1.5 changes per minute. This indicates that participants experience connection issues with the networking infrastructure.

4.4 P2P Transfers

Frequent dis- and reconnects trigger the protocol switch more often, which may increase the number of aborted p2p transfers. A p2p transfer is aborted in two cases: First, while the transfer is in progress, the remote end loses network connectivity and closes all its serving sockets to inform its clients about the change allowing them to find a new server peer. The client peer records this as an aborted transfer. Second, the device reconnects to the network, stops the p2p transfer, and uploads data itself rather than risking a disconnect from the remote end.

Figure 7 depicts the average count of completed and aborted transfers over time. The analysis focuses on the two most prominent cases which repeat themselves throughout but possibly to a lesser extent. It shows an increased number of aborts from 13:00 onwards, which means that more transfers are affected by the connectivity change. In contrast, the transfers at noon (11:25 to 13:00) are less affected since the number of completed transfers is higher than the aborted ones. One reason is that at noon some devices experience disconnects that are much longer than induced by the software controller. Whether they are in a blind spot or deliberately turned their networking interfaces off, is unclear. Apart from that, not every connectivity change happens during a p2p data transfer but instead to peers that are currently idle.

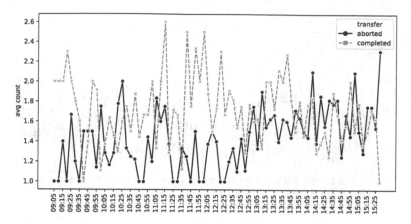

Fig. 7. At around noon, p2p transfers are much less affected by connectivity changes than from 13:00 onwards when the number of aborted transfers increases.

4.5 Data Delay

Data delay is the time from a senor reading being collected to when it becomes available at the server. The analysis refers to the mock sensor readings, which have been uploaded to the server either directly or via a peer. For completed transfers, the data delay is well below 200 s (cp. Fig. 8a). The mean delay is around 100 s. Given the small amounts of data transferred, this delay is rather high. This is caused by the timeout between protocol iterations. If a device did not discover a connected peer, it idled for 60 s before it run the p2p part of the protocol again.

For data directly uploaded by the device itself, the mean delay is at 25 s, which is expected (cp. Fig. 8b). Notice, however, the density of outliers, which show that a considerable number of devices deviate from the average. One explanation may be that these devices started out with a p2p transfer, regained network connectivity, aborted the transfer, and uploaded data themselves. This takes longer than uploading data right away. There are, however outliers whose delay is too long to fit into the same graph (cp. Fig. 8c). In these cases, the devices remained disconnected for a long time, during which they were unable to find a connected peer in their proximity. Or, during which the payload had grown to a size that no peer connection was stable enough to complete the transfer, either because the peer moved out of range, or the peer lost connectivity. Generally, the phone creates 84 Bytes of mock readings per second adding up to 300 Kilobytes an hour.

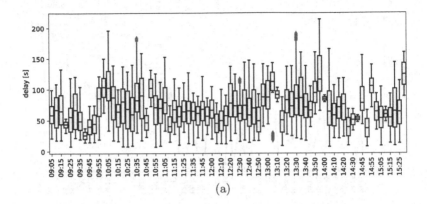

(a)

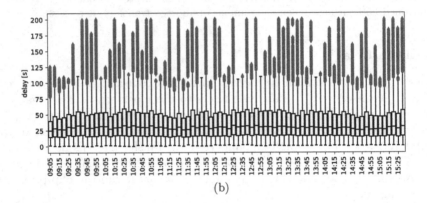

(b)

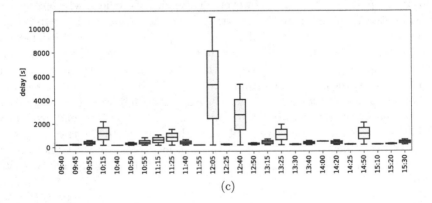

(c)

Fig. 8. The delay for completed transfers and subsequent uploads would be rather swift if it was not for the timeout between protocol iterations (a). For direct uploads a considerable number of delays deviate from the otherwise short average delay (b). Some direct uploads take so long that they are outsourced to a separate graph (c).

5   Conclusion

The availability of crowd monitoring data is essential for event organisers and emergency response personnel to ensure safety. In case of network issues, data transfers to peers in vicinity that are still connected may allow for sharing such data nonetheless. This paper focuses on the delivery of crowd monitoring data. During an experiment at an indoor fair, a p2p transfer protocol was tested for how it meets real-world challenges. With considerable effort put into the design of the events' companion app, 47 visitors agreed to become visible to peers and share sensor and log data. The results of the experiment show:

- Despite variations in the continuity and duration of participation, most participants were part of the same large radio cluster throughout the experiment and thus fulfilled the main prerequisite for p2p data transfers.
- Even at this medium size indoor event, network issues occur when the infrastructure is not particularly adjusted to the expected number of visitors. This underlines the necessity for solutions that mitigate the loss of data.

Further, unanticipated changes in the connectivity to the network infrastructure challenged the protocol and highlight prospective refinements of its design:

- Participants experienced at times high rates of connectivity changes. The protocol quickly switched between dis- and reconnects often aborting p2p data transfers. This is inevitable when the remote end cancels the connection. However, losing the network connection could be approached more gracefully: Instead of immediately switching to a p2p transfer, a device could wait to see if it can reconnect after a short time. When the disconnect does take longer, a p2p transfer is feasible as the experiment shows short delays when a connected peer is around.
- Disconnects that lasted dozens of minutes caused the steady growth of data to be transferred. Agnostic of these changes, the protocol kept trying to transfer all collected data at once. For this to succeed, a peer connection would be required to be stable for an extended period of time. A more realistic way to mitigate this built-up of data is to split it up into small chunks that are suitable for short connections times with peers.

Overall, the lesson learnt from testing the protocol in a real-world setting is that attention to detail is important before advancing the protocol design e.g., to a multi-hop solution.

References

1. Blanke, U., Tröster, G., Franke, T., Lukowicz, P.: Capturing crowd dynamics at large scale events using participatory GPS-localization. In: IEEE International Conference on Intelligent Sensors, Sensor Networks and Information Processing, Conference Proceedings (ISSNIP), pp. 21–24. IEEE (2014). https://doi.org/10.1109/ISSNIP.2014.6827652

2. Castagno, P., Mancuso, V., Sereno, M., Marsan, M.A.: Why your smartphone doesn't work in very crowded environments. In: IEEE 18th International Symposium on A World of Wireless, Mobile and Multimedia Networks (WoWMoM), pp. 1–9, June 2017. https://doi.org/10.1109/WoWMoM.2017.7974296
3. Danielis, P., Kouyoumdjieva, S.T., Karlsson, G.: Urbancount: mobile crowd counting in urban environments. In: 2017 8th IEEE Annual Information Technology, Electronics and Mobile Communication Conference (IEMCON), pp. 640–648, October 2017. https://doi.org/10.1109/IEMCON.2017.8117189
4. Groba, C., Springer, T.: Exploring data forwarding with bluetooth for participatory crowd monitoring. In: 2019 IEEE International Conference on Pervasive Computing and Communications Workshops (PerCom Workshops), pp. 71–76, March 2019. https://doi.org/10.1109/PERCOMW.2019.8730711
5. Kannan, P.G., Venkatagiri, S.P., Chan, M.C., Ananda, A.L., Peh, L.S.: Low cost crowd counting using audio tones. In: Proceedings of the 10th ACM Conference on Embedded Network Sensor Systems, SenSys 2012, pp. 155–168. Association for Computing Machinery, New York (2012). https://doi.org/10.1145/2426656.2426673
6. Kjægaard, M.B., Wirz, M., Roggen, D., Tröster, G.: Mobile sensing of pedestrian flocks in indoor environments using wifi signals. In: 2012 IEEE International Conference on Pervasive Computing and Communications, pp. 95–102, March 2012. https://doi.org/10.1109/PerCom.2012.6199854
7. Kluge, T., Groba, C., Springer, T.: Trilateration, fingerprinting, and centroid: taking indoor positioning with bluetooth LE to the wild. In: 21st International Symposium on "A World of Wireless, Mobile and Multimedia Networks" (WoWMoM) (WoWMoM 2020). Cork, Ireland, June 2020. Accepted
8. Loomba, R., de Frein, R., Jennings, B.: Selecting energy efficient cluster-head trajectories for collaborative mobile sensing. In: IEEE Global Communications Conference (GLOBECOM), pp. 1–7 (2015). https://doi.org/10.1109/GLOCOM.2015.7417727
9. Wirz, M., Franke, T., Roggen, D., Mitleton-Kelly, E., Lukowicz, P., Tröster, G.: Probing crowd density through smartphones in city-scale mass gatherings. EPJ Data Sci. 2(1), 1–24 (2013). https://doi.org/10.1140/epjds17
10. Wirz, M., Schläpfer, P., Kjundefinedrgaard, M.B., Roggen, D., Feese, S., Tröster, G.: Towards an online detection of pedestrian flocks in urban canyons by smoothed spatio-temporal clustering of GPS trajectories. In: Proceedings of the 3rd ACM SIGSPATIAL International Workshop on Location-Based Social Networks. LBSN 2011, p. 17–24. Association for Computing Machinery, New York (2011). https://doi.org/10.1145/2063212.2063220
11. Zhang, J., Guo, H., Liu, J.: Energy-aware task offloading for ultra-dense edge computing. In: 2018 IEEE International Conference on Internet of Things (iThings) and IEEE Green Computing and Communications (GreenCom) and IEEE Cyber, Physical and Social Computing (CPSCom) and IEEE Smart Data (SmartData), pp. 720–727 (2018). https://doi.org/10.1109/Cybermatics_2018.2018.00144

Reliability Enhancement of URLLC Traffic in 5G Cellular Networks

Jerzy Martyna[✉]

Institute of Computer Science, Faculty of Mathematics and Computer Science, Jagiellonian University, ul. Prof. S. Lojasiewicza 6, 30-348 Cracow, Poland
jerzy.martyna@uj.edu.pl

Abstract. 5G cellular networks must be able to deliver a small data payload in a very short time (up to 1 ms) with ultra-high probability of success (99.999%) to the mobile user. Achieving ultra-reliable and low-latency communication (URLLC) represents one of the major challenges in terms of system design. This paper covers definitions of latency and the reliability of URLLC traffic. Furthermore, it presents a method for reliability enhancement of URLLC traffic. To this end, the problem of reliability enhancement is formulated as an optimisation problem, the objective of which is to maximise the sum of data rates for all users with the URLLC constraints. Simulation results show that the suggested method validates the proposed model.

Keywords: 5G systems · Wireless scheduling · URLLC traffic · eMBB traffic · Reliability

1 Introduction

The newly introduced fifth generation (5G) mobile cellular network is the first wireless network standard designed to support multi-service communication [1]. More specifically, 5G aims to cover three generic connectivity types: enhanced Mobile Broadband (eMBB), massive Machine-Type Communication (mMTC) and Ultra-Reliable Low-Latency Communication (URLLC). eMBB is an enhancement of the mobile broadband services of the current long term evolution (LTE) system. mMTC service provides massive connectivity solutions for various Internet of Things (IoT) applications. The main design goals are supporting high density of devices (up to a million devices per square kilometer) and significant extension of the lifetime of individual devices (up to 10 years battery lifetime) [2]. Ultra-Reliability Low-Latency Communication (URLLC) supports low latency transmissions (0.25–0.3 ms/packet) with high reliability (99.999%) [3]. Satisfying these very high requirements makes 5G network implementation a major design challenge.

In general, in 5th generation networks, the problem of three traffics can be treated by two approaches. The first approach involves analyzing orthogonal slicing. Then all different slices are allocated to all three types of traffics.

© Springer Nature Switzerland AG 2020
P. Gaj et al. (Eds.): CN 2020, CCIS 1231, pp. 77–88, 2020.
https://doi.org/10.1007/978-3-030-50719-0_7

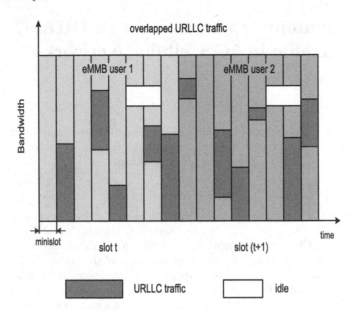

Fig. 1. Superposition approach for multiplexing eMMB and URLLC in 5G cellular network.

Then there is no interference between these traffics. This approach is characterized by inefficiency in the use of resources. It was analyzed, among others in papers [4–6]. Unlike previously described, the second approach does not require orthogonal slicing. The use of non-orthogonal slicing allows you to increase resource efficiency, but can cause interference between these traffics. This approach was used, among others, in [7–9].

This work considers non-orthogonal slicing for downlink resource allocation of URLLC and eMBB. The third type (mMTTC) support a large number of things (IoT) devices and can be used sporadically, so this will not be considered here either. In this work it was assumed that each time slot is divided into 0.125 ms minislots [2,10]. As can be seen in Fig. 1, this describes this structure, within each slot, eMBB traffic can share the bandwidth over the time-frequency plane. Time is divided into slots, and further is also divided into minislots. eMBB traffic is located at the beginning of the slots, while URLLC traffic can be overlapped at any minislot. A number of users are flexibly multiplexed over the available resources with different transmission time interval (TTI) durations. The TTI size can be dynamically adjusted according to the number of users, their requirements, etc. A long TTI allows us to take advantage of the benefits of coding gains closer to the Shannon capacity limit, and also imposes a lower control overhead. Unfortunately, this can cause an increase in latency. Hence, it is obvious that a proper scheduling of users can minimise the latency and reliability requirements.

Message transmission for mission-critical applications needs a latency of not more than a few milliseconds or even lower than 1 ms for fully autonomous applications [11]. It is not possible to use the LTE system here, because it provides a delay of 30 to 100 ms and this is unacceptable. The opposite is provided by a special URLLC service described a.o. by P. Schulz, *et al.* [12]. To achieve this, the grant-free transmission mechanism of the physical layer access was proposed by C. She *et al.* [13]. The appropriate modulation schemes and the number of links are proposed as a solution to this problem by W. Anwar *et al.* [14]. In turn, the multi-connectivity activation scheme for URLLC constraints as a new approach in their implementation was proposed by J. Rao *et al.* [15].

In this paper, the problem of reliability enhancement in the finite blocklength regime subject in URLLC traffic is discussed. Through the optimisation of the data rate, with the reliability constraints it is possible to obtain all the required parameters of the URLLC downlink packets arriving during an eMBB transmission. This allows us to find a solution that is different from those obtained by minimising just the mean resource utilisation. In addition, the contribution of this article is the development of a new online approximated algorithm that allows you to increase the reliability of URLLC traffic regardless of the load.

The remaining parts of this paper are as follows. Section 2 presents the system model. In this section the problem of the reliability enhancement is formulated as an optimisation problem with reliability constraints and transformed into deterministic form. Section 3 presents the optimal resource allocation problem. In Sect. 4, the online heuristic algorithm for reliability enhancement of URLLC traffic is provided. Simulation results are then presented in Sect. 5, followed by concluding remarks in Sect. 6.

2 System Model

2.1 Traffic Model

In this model, the time is divided into equally spaced slots with one millisecond time duration, which is compatible with current cellular network solutions. The downlink eMBB traffic originating from the backlogged users shares the bandwith over the time frequency plane in each slot and is fixed during that slot. The downlink stochastic URLLC traffic may arrive during the time slot which is allocated to different eMBB users. The URLLC traffic cannot be queued until the next slot. Therefore, each eMBB slot is divided into minislots, each of which has a 0.125 ms duration. This means that each arrived URLLC traffic is scheduled immediately in the next minislot on top of the ongoing eMBB transmission (see Fig. 1).

2.2 The Reliability Requirements of URLLC Traffic

The system bandwidth W is chosen that the probability of blocking of a URLLC packet arrival is of the order of δ. The QoS parameters of URLLC traffic d and δ

are specified as follows: a packet must be successfully delivered within a end-to-end delay of no more than d seconds with a probability of at least $1 - \delta$. Thus, δ means here the reliability of the URLLC traffic. Let λ be the system load and $\mathbf{r} = (r_1, r_2, \ldots, r_C)$ be a number of channels, where C is total number of classes. Each class represents users with the same SINR. Let the set of all classes be given as $\mathcal{C}$. Thus, the following condition must be satisfied, namely [16,17]:

$$W \geq \zeta^{mean}(\mathbf{r}) + c(\delta)\sqrt{\zeta^{variance}(\mathbf{r})} \tag{1}$$

where $c(\delta) = Q^{-1}(\delta)$, $Q(.)$ is the Q-function, $\zeta^{mean}(\mathbf{r}) = \sum_{c=1}^{C} \lambda_c \frac{r_c}{\kappa}$ is the mean bandwith utilisation and $\zeta^{variance}(\mathbf{r}) = \sum_{c=1}^{C} \frac{r_c^2}{\kappa^2 d}$ is the variance of the bandwith utilisation, κ is the a constant which denotes the number of channel uses per unit time per unit bandwidth of the OFDMA time-frequency plane.

2.3 Joint eMBB/URLLC Scheduling in One Slot

The scheduling combines two movements dependent on the eMBB state and the URLLC traffic, which is a strategy for placement across minislots. This strategy takes into account the eMBB users, and in turn the URLLC must be located so that their requests or blocking are included. Therefore, to carry out this scheduling, the URLLC traffic data should be allocated in each minislot, if one is required. This is done by affecting the data rate of eMBB traffic. Thus, the data rate of the m-th eMBB user is given as follows:

$$R^m_{eMBB} = \sum_{i=1}^{N}(b_i - f_{m,i})\log_2(1 + SINR_i) \tag{2}$$

where b_i is the resource allocated to URLLC user i, $f_{m,i}$ is the busy resource of eMBB m-th user by the URLLC data, $SINR_i$ is the signal-to-noise ratio of i-th URLLC user, N is the total number of URLLC user.

The data rate of the i-th URLLC user on subcarrier k is given by

$$R^i_{URRLC} = \log_2(1 + p_{i,k}\gamma_{i,k}) \tag{3}$$

where $p_{i,k}$ is the transmission power to the i-th URLLC user on the subcarrier k, $\gamma_{i,k} = h_{i,k}/(N_0 W + I_{i,k})$ is the signal to interference plus noise ratio (SINR), $h_{i,k}$ is the channel gain on subcarrier k and the i-th URLLC, N_0 is the noise power and $I_{i,k}$ is the interference introduced to the i-th URLLC user on the subcarrier k.

The total data rate in a downlink transmission is given by

$$R = \sum_{m=1}^{M} R^m_{eMMB} + \sum_{i=1}^{N} R^i_{URLLC} = \frac{W}{\Theta} \sum_{j=1}^{M+N} \sum_{i=1}^{\Theta} \log_2(1 + \frac{\gamma_{j,i}}{\Gamma}) \tag{4}$$

where Γ is a function of the required bit-error rate (BER) and is approximately equal $\Gamma \overset{\triangle}{=} -\ln(5BER)$ [18]. In the range of BER $< (\frac{1}{5})\exp(-1.5) \approx 0.0446$. Θ is the number of orthogonal subbands.

3   Optimal Resource Allocation

This section presents the problem of optimal resource allocation of the URLL data in the eMBB data traffic with the reliability enhancement.

According to Eq. (2), the achievable eMBB data rate at each time T is given by

$$\max \sum_{t=1}^{T} \sum_{i=1}^{N} \sum_{m=1}^{M} w_{i,m}^{(t)} R_{eMMB}^{(m,t)} \tag{5}$$

subject to:

$$C1: \quad R_{eMBB}^{(m,t)} \geq R_{eMBB}^{req}, \quad \forall m \in \mathcal{M}, \quad \forall t \in \{1, \ldots, T\} \tag{6}$$

$$C2: \quad \sum_{r \in C} \left(\sum_{m \in \mathcal{M}} R_{eMBB}^{(m,t)} x_{m,r}^{(t)} + \sum_{i \in \mathcal{N}} R_{URLLC}^{(i,t)} x_{i,r}^{(t)} \right) \leq R, \quad x_{m,r}^{(t)} \in \{0,1\},$$

$$x_{i,r}^{(t)} \in \{0,1\}. \quad \forall r \in \{r_1, r_2, \ldots, r_C\}, t \in \{1, \ldots, T\} \tag{7}$$

$$C3: \quad x_{m,r}^{(t)} + x_{i,r}^{(t)} \leq 1, \quad x_{m,r}^{(t)} \in \{0,1\}, \quad x_{i,r}^{(t)} \in \{0,1\}, \quad m \in \mathcal{M}, \quad i \in \mathcal{N},$$

$$\forall r \in \{r_1, r_2, \ldots, r_C\}, \quad t \in \{1, \ldots, T\} \tag{8}$$

$$C4: \quad R_{URLLC}^{(i,t)} \leq R_{URLLC}^{res}, \quad \forall i \in \mathcal{N}, \quad \forall t \in \{1, \ldots, T\} \tag{9}$$

where the $C1$ condition represents a limitation of required data rate for each m-th eMBB traffic user in slot t; the $C2$ condition is a limitation of the system data rate for all M eMBB traffic users and N URLLC users in the slot t. Equation $C3$ ensure that the r-th channel will be utilised in the slot t. The $C4$ condition represents the data rate restriction that can be used by i-th URLLC traffic user in the slot t. $w_{i,m}^{(t)}$ is the weight in the t-th slot for m-th eMBB user and i-th URLLC traffic user in the slot t. $x_{m,r}^{(t)}$ denotes the traffic indicator, i.e. $x_{m,r}^{(t)} = 1$ only if the m-th eMBB traffic is allocated to the r-th channel and $x_{m,r}(t) = 0$ in the slot t otherwise. Similarly, $x_{i,r}^{(t)} = 1$ indicated that the r-th channel is assigned to the i-th URLLC user in the slot t and $x_{n,r}^{(t)} = 0$ otherwise.

The optimisation problem is stochastic, nonlinear, non-convex and includes three variables, namely binary $x_{m,r}^{(t)}$, $x_{i,r}^{(t)}$ and $w_{i,m}^{(t)}$. So, the problem giving in Eq. (5) is a Stochastic Mixed Integer Programming (SIMP) [19]. There are two elements to this modelling, namely: Stochastic Programming (SP) and Mixed-Integer Programming (MIP). Each element is important to capturing the different factors involved in the problem. The computational complexity of finding the exact solution is very high thus not reasonable in practice. The alternative is to use heuristic approaches. In Sect. 4, a novel polynomial time heuristic algorithm for the suboptimal traffic scheduling for reliability enhancement is proposed.

4   Online Heuristic Algorithm for Reliability Enhancement of URLLC Traffic

The proposed online heuristic algorithm is based on sliding windows model [20]. This model allows to perform the required computations using the stream generated by a single scan of the data. This dynamic approach provides less precise statements than the static one since it uses less information and it has higher implementation overhead. However, this model allows progressive creation of the planning sequence.

Each scheduling period in the presented model corresponds here to the assumed window length, which is the slot time. Packets belonging to the both traffics, namely eMBB and URLLC, are scheduled in this window. These packets in the t-th scheduling period form a set, namely: $\{SP\}_t = \{packet_{t,1}, \ldots, packet_{t,J}\}$, where J is the total number of packets in time slot. The packet scheduling is based on the arrival time, the required end-to-end delay, the reliability, the waiting time, and the packet usability. All of the packet parameters listed above have been standardised according to the following relationship:

$$f(y_k) = \frac{y_k - \min_{packet \in \{SP\}_t}(y_k) + 1}{\max_{packet \in \{SP\}_t}(y_k) - \min_{packet \in \{SP\}_t}(y_k) + 1} \tag{10}$$

Four priorities have been calculated for such standardised parameters. So, priority of the time arrival for j-th packet is given by

$$A_j = \frac{AP_j}{UP_j} \tag{11}$$

where AP_j is the standarised time arrival of packet j and UP_j is the standarised usability of packet j. The priority of the required end-to-end delay for the packet j is as follows

$$D_j = \frac{DP_j}{UP_j} \tag{12}$$

where DP_j is the standarised end-to-end delay of packet j. The priority of the reliability for j-th packet is defined as follows

$$R_j = \frac{RP_j}{UP_j} \tag{13}$$

where RP_j is the standarised reliability of packet j. The usability of the packet j is given by

$$U_j = \frac{UP_j}{TT_j} \tag{14}$$

where TT_j is type of traffic to which j-th packet belongs. It is assumed here that the parameter $TT = 1$ and $TT = 2$ for eMBB traffic and URLLC traffic, respectively.

In addition, to avoid possible conflicts between packages, resulting from, among others due to lack of space in the mini-slot, a conflict index γ is entered for each packet j. It is defined as the number of packets in conflict with the j-th packet. Then for each packet j can be specified the degree of conflict DC_j as follows:

$$DC_j = \frac{Z_j}{(1 + \gamma_j)^2} \tag{15}$$

where Z_j is the number of packets that are observed by the packet j and are in potential conflict with packet j. If the DC_j value is greater than the set value, the packet j is suspended and placed at the beginning of a new slot.

The concept of the heuristic algorithm is as follows:

1) The number of packets that can be placed in a single window is found initially. Packets from the buffer that are not included in the previous window are attached to these packets. For each packet separately all parameters are calculated, namely: A_j, D_j, R_j, U_j.

2) A value DC_j is also calculated for each packet to be placed in a single window. If the DC_j value exceeds the set bound, then the j-th packet with this value is removed from the list of packets to be placed in the window. This packet is placed in the packet buffer, which will be placed in the next window.

4) Parameters such as W, R_{eMBB}, ρ, δ are calculated for the entire list of packets to be placed in the window. If these parameters are not satisfied, all holes in the minislots are filled with packets from the buffer.

5) If the entire contents of the list of packets found for the window is accepted, then all of them are accepted for transmission in the window.

The pseudocode of the algorithm for placing packets in the sliding window is represented by Algorithm 1. It consists of two procedures: *Sliding window* and *Heuristic*. The first one prepares a list of $\{SP\}$ packets for each window t, for which all necessary parameters are calculated. Then sorts them by priority. All sorted packets are placed in the $\{SP\}$ list, which contains packets to be placed in the window. But before their final acceptance, the *Heuristic* procedure is called. Its purpose is to check the degree of possible conflicts. If their values are too high, it removes the packet from the $\{SP\}$ list and places it in the buffer. Then, until the performance parameters $(W, R_{eMBB}, \rho, \delta)$ for this scheduling are met, it will check for any free spaces in each minislot l. If he finds them, he fills them with packets from the buffer. Packets that meet all requirements are sent in the window by *Send_to_transmission* procedure.

The proposed algorithm has the complexity of $O(J^2)$ per slot, where J is the total number of packets per slot and scheduled in a single window (Table 1).

Algorithm 1. Heuristic online algorithm for reliability enhancement

1: **procedure** SLIDING WINDOW
2: **Require:** $\{buffer\}, \{SP\}, \{Final\_SP\}$
3: **Initialisation:**
4:     Let $t \leftarrow 1$;
5:     Let $\{buffer\} \leftarrow \varnothing$;
6:     Let $\{SP\}_t \leftarrow \varnothing$;
7:     Let $\{Final\_SP\} \leftarrow \varnothing$;
8:     **for** $t \leftarrow 1, T$ **do**
9:         $j \leftarrow 1$
10:        **if** $\{buffer\} \neq \varnothing$ **then**
11:            Copy $\{buffer\}$ to $\{SP\}_t$
12:            $\{buffer\} \leftarrow \varnothing$
13:        **end if**
14:        **repeat**
15:            Add $packet_j$ to $\{SP\}_t$
16:            Calculate $A_j, D_j, R_j, U_j, \gamma_j$
17:        **until** $j > J$
18:        Sort all packets in $\{SP\}_t$
19:        Heuristic($\{SP\}_t, \{Final\_SP\}$)
20:        Send_to_transmission($\{Final\_SP\}$)
21:        $\{SP\}_t \leftarrow \varnothing$
22:        $\{Final\_SP\} \leftarrow \varnothing$
23:    **end for**
24: **end procedure**

25: **procedure** HEURISTIC($\{SP_t\}, \{Final\_SP\}$)
26: **Require:** $minislot[L]$
27:    **while** $\{SP\} \neq \varnothing$ **do**
28:        **repeat**
29:            Calculate DC_j
30:            **if** $DC_j > Limit$ **then**
31:                Copy $packet_j$ to $buffer$
32:            **end if**
33:        **until** $j > J$
34:        **while** $(W_J < W_{req}$ or $R^J_{eMBB} \geq R^{req}_{eMBB}$ or $\delta < \delta^{req}$ or $\rho < \rho^{req})$ **do**
35:            $l \leftarrow 1$
36:            **repeat**
37:                Find free holes in $minislot_l$
38:                Copy packets from the $buffer$ to $minislot_l$
39:            **until** $l > L$
40:        **end while**
41:        Copy $\{SP\}_t$ to $\{Final\_SP\}$
42:    **end while**
43: **end procedure**

Table 1. Main parameters of simulation

Parameter	Value
Source image size	800×600
IFFT size	512
Number of OFDM subcarriers	256
OFDM symbol length [μs]	60
OFDM symbol number	4
Carrier frequenxy [MHz]	2000
Modulation method	QPSK
CRC Length [bits]	16
Propagation conditions	Log-Normal shadowing with 5 dB standard deviation; $128.1 + 37.6 \log(D[km])$
Signal-to-Noise Ratio (SNR)	5 dB
Transmission bandwith	20 MHz
Network environment	17 picocells
Traffic elements	URLLC 10 users/cell; eMBB 10 users/cell

5   Simulation Results

In this section, the results of simulations of the proposed resource allocation scheme for reliability enhancement are presented. The simulation environment comprises a multicell 5G network constructed according to the network model in Sect. 3, where multiple picocells share the spectrum resources with one macrocell and a variable number of D2D connections.

Firstly, the scheduling to ensure eMBB throughput was checked. It has been assumed that 10 eMBB users per picocell are active in the system. The eMBB slot of each of these users is composed of eight minislots. For seven eMBB users, the probability of distribution was chosen so that the average rate is equal 5 Mbps, and for the other three probability distribution is such that the average rate is 3 Mbps. The number of listed states here is around 1 million. URLLC traffic is scheduled in minislots so that peak URLLC load in an eMBB slot is less than or equal to $1 - \delta$.

Figure 2 shows the average rate of both traffics versus the average rate of URLLC load for two scheduling policies: the proportional-fair sharing algorithm [21] for eMBB users and the heuristic online algorithm. The graph shows that the system with the heuristic algorithm is characterised by a greater average data rate than the standard eMBB algorithm. This is especially evident in the case of high average URLLC load.

Then the system reliability obtained was tested. Figure 3 presents the sum data rates of eMBB users versus the reliability levels of URLLC traffic for both scheduling policies. The figure shows that the use of the heuristic online algorithm provides a higher sum of data rate for eMBB, which means that this algorithm improves data flow at the same URLLC reliability levels.

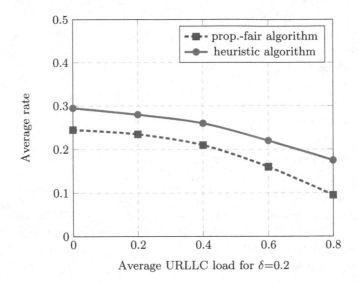

Fig. 2. Average rate of eMBB and URLLC traffic versus average URLLC load for the proportional-fair sharing algorithm and the heuristic online algorithm.

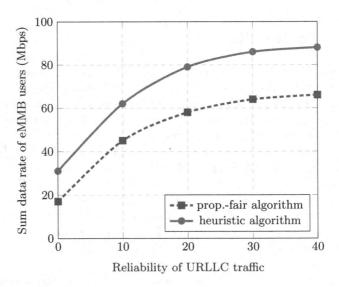

Fig. 3. Sum data rates of eMBB users versus the reliability levels of URLLC traffic for the proportional-fair sharing algorithm and the heuristic online algorithm.

Finally, the reliability levels at 1 ms latency for different URLLC loads was studied for both scheduling policies. Figure 4 shows the reliability at 1 ms latency versus different URLLC loads for both scheduling policies. It can be seen from the figure that the use of the proposed scheduling algorithm allows reliability to be increased by approx. 15–20%, regardless of the URLLC traffic load.

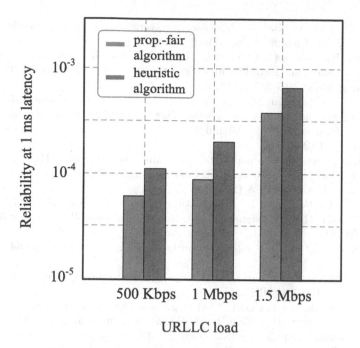

Fig. 4. The reliability at 1 ms latency versus average load of URLLC for the proportional-fair sharing algorithm and the heuristic online algorithm.

6 Conclusion

In this paper, the problem of reliability enhancement in the 5G cellular network was studied. An optimal resource allocation policy is proposed to maximise the reliability. Via simulation, the achievable reliability of the network with the proposed resource allocation policy was used. In particular, the achievable reliability obtained using standard methods was compared with the reliability obtained by applying the proposed solution of maximising the reliability in the studied model. Significant performance differences between these reliabilities are observed, which confirm the necessity and contribution of this paper.

References

1. ITU-R M.2083-0, IMT Vision - Framework and overall objectives of the future development of IMT for 2020 and beyond, September 2015
2. 3GPP TSG RAN WG1 Meeting 87, November 2016
3. 3GPP TR 38.913 V14.2.0, 5G; Study on Scenarios and Requirements for Next Generation Access Technologies (2017)
4. Zhang, L., Ijaz, A., Xiao, P., Quddus, A., Tafazolli, R.: Subband filtered multi-carrier systems for multi-service wireless communications. IEEE Trans. Wireless Comm. **16**(3), 1893–1907 (2017)

5. Zhang, L., Ijaz, A., Xiao, P.: Multi-service system: an enabler of flexible 5G air-interface. IEEE Commun. Mag. **55**(10), 152–159 (2017)
6. Pedersen, K., Pocovi, G., Steiner, J., Maeder, A.: Agile 5G scheduler for improved E2E performance and flexibility for different network implementations. IEEE Commun. Mag. **56**(3), 210–217 (2018)
7. Kowalski J. M., Nogami T., Yin Z., Sheng J., Ying K.: Coexistence of enhanced mobile broadband communications and ultra reliable low latency communications in mobile fronthaul. In: Broadband Access Communication Technologies XII, no. January, p. 11 (2018)
8. Anand, A., Veciana, G., Shakkottai, S.: Joint scheduling of URLLC and eMBB traffic in 5G wireless networks. IEEE International Conference on Computing Communication, Honolulu, USA (2018)
9. Esswie, A.A., Pedersen, K.I.: Opportunistic spatial preemptive scheduling for URLLC and eMBB coexistence in multi-user 5G networks. IEEE Access **6**, 38451–38463 (2018)
10. Hoymann, C., et al.: LTE release 14 outlook. IEEE Commun. Mag. **54**(6), 44–49 (2016)
11. ITU-T: The Tactile Internet (2014). https://www.itu.int/dms_pub/itu-t/oth/23/01/T23010000230001PDFE.pdf
12. Schulz, P., et al.: Latency critical IoT application in 5G: perspective on the design of radio access networks. IEEE Trans. Wirel. Commun. **55**(2), 70–78 (2017)
13. She, C., Yang, C., Quek, T.S.: Cross-layer optimization for ultra-reliable and low-latency radio access networks. IEEE Trans. Wirel. Commun. **17**(1), 127–141 (2018)
14. Anwar, W., Kulkarni, K., Franchi, N., Fettweis, G.: Physical layer abstraction for ultra-reliable communications in 5G multi-connectivity networks. In: IEEE Annual International Symposium on Personal, Indoor and Mobile Radio Communications (PIMRC), Italy, Bologna (2018)
15. Rao, J., Vrzic, S.: Packet duplication for URLLC in 5G dual connectivity architecture. In: 2018 IEEE Wireless Communications and Networking Conference (WCNC), Barcelona, Spain, April 2018
16. Harchol-Balter, M.: Performance Modeling and Design of Computer System: Queueing Theory in Action. Cambridge University Press, Cambridge (2013)
17. Anand, A., de Veciana, G.: Resource allocation and HARQ optimization for URLLC traffic in 5G wireless networks. http://arxiv.org/abs/1804.09201
18. Jang, J., Lee, K.B.: Transmit power adaptation for multiuser OFDM systems. IEEE J. Sel. Areas Commun. **21**(2), 171–178 (2003)
19. Burge, J.R., Louveaux, F.V.: Introduction to Stochastic Programming. Springer, New York (1997). https://doi.org/10.1007/b97617
20. Datar, M., Gionis, A., Indyk, P., Motwani, R.: Maintaining stream statistics over sliding windows. SIAM J. Comput. **31**(6), 1794–1813 (2002)
21. Kushner, H.J., Whiting, P.A.: Convergence of proportional-fair sharing algorithms under general conditions. IEEE Trans. Wirel. Commun. **3**(4), 1250–1259 (2004)

Cybersecurity and Quality of Service

Detection Efficiency Improvement in Multi–component Anti-spam Systems

Tomas Sochor[✉]

Faculty of Science, Department of Informatics and Computers,
University of Ostrava, 30 dubna 22, 70103 Ostrava, Czechia
tomas.sochor@osu.cz
http://www1.osu.cz/home/sochor

Abstract. Multi–layer spam detection systems frequently used in many SMTP servers often suffer from a lack of mutual communication between individual layers. The paper presents the construction of a feedback interconnection between two significant layers, namely Message contents check and Greylisting. The verification in a real SMTP server is performed, demonstrating considerable improvement of spam detection efficiency comparing the previous period with missing interconnection, while for a short testing period. Despite the limited generalizability of the result, it suggests the easy way how spam detection can be improved.

Keywords: Spam detection · Multi–layer detection · Blacklisting · Greylisting · Message contents check · SMTP dialog

1 Introduction

Unwanted (unsolicited) messages in electronic mailboxes (called spam) are known virtually to every e-mail user. The amount of spam is so high that without efficient spam detection filter the electronic mail service renders to be useless because about 80% of the total number of e-mail messages sent or more are spam messages according to [1].

2 Objectives

There are many different approaches to detect and eliminate spam. In the first phase of spam detection and elimination spanning from the late 1970s till the early 1990s, the main focus was given on looking into the contents of individual messages as delivered into the e-mail client and classification spam according to the presence of predefined fixed characters. Nevertheless, towards the end of the phase, it became apparent that more sophisticated detection mechanisms are necessary.

First, the main focus of spam detection moved from end–stations to SMTP servers. As a result, numerous collaborative detection techniques like collaborative filtering [2] came into the limelight. Also, the spam reception avoiding

© Springer Nature Switzerland AG 2020
P. Gaj et al. (Eds.): CN 2020, CCIS 1231, pp. 91–100, 2020.
https://doi.org/10.1007/978-3-030-50719-0_8

methods like blacklisting and greylisting making use of an SMTP–specific feature called SMTP dialog sending the message metadata (namely sender and recipient e-mail addresses) were widely applied [3,4]. Since then, the spam detection happens at the point of entry into a corporate/organization network, i.e., on an SMTP server of the organization [5,6]. Nevertheless, none single among them can eliminate every spam message or even a sufficient amount hereof as documented, e.g. in [7].

2.1 Multi–layer Spam Detection

Despite recent tendencies to outsource the spam detection tasks to third–party cloud providers, there are still many e-mail servers performing the spam detection on their own. There, multicomponent (multi–layer) spam detection systems are used frequently. A typical multi-layer spam detection system consists of the following:

- Initial formal check,
- Blacklisting,
- Greylisting,
- Message contents check.

Also other components may appear, the list is not intended to be exhaustive. The component order can vary, but Blacklisting and Greylisting are usually performed before the message contents are checked. The reason for Blacklisting and Greylisting ordering first is that these components can prevent a significant part of spam messages from being delivered as demonstrated, e.g. in [8], and incorporated into the broader security framework in [9,10].

The detection layers in the list above tend to be implemented as separate software components. Thus, their interaction can be minimized. The positive aspect consists in the elimination of interlocks, e.g., because of software failures. On the other hand, the complete insulation of the component can affect their detection efficiency as it was shown, e.g. in [11–13].

3   Methods

The aim of the research presented in this paper was to design an interconnection between specific components of a general multi–layer detection system and perform a verification of the effect in a real environment.

The first question is what components need to be interconnected. This question must be answered not only because a full-mesh interconnection could be overly complicated, but primarily regarding the potential benefit of the proposed interconnection. As it was analyzed in details in [11], the significant improvement can be expected from the interconnection between Greylisting (as a component detecting spam based on somewhat limited information from SMTP dialog) and the Message contents check (this component has complete information about the message, and the spam detection reliability here is much higher).

In most Greylisting implementations, not all message delivery attempts are subject to the spam check (consisting in temporary delivery refusal). In addition to manually whitelisted senders, also messages from senders that appear to be legitimate regarding their previous behavior are excluded (not checked). Usually, such senders (the IP addresses of sending servers) are listed in the so-called Automatic WhiteList (AWL). The exclusion of some delivery attempts is motivated by decreasing the system resources necessary for the Greylisting spam check.

For AWL enlisting, it is enough when the SMTP server successfully delivers several (e.g., 5) messaged recently. Unfortunately, this can also be done by more sophisticated spamming software. This drawback could be overcome if the Greylisting module is informed that a message originated from a specific AWL-listed SMTP server was classified as spam by the following Message contents check. Therefore, just the feedback interconnection between Message contents check and Greylisting was designed and later verified in the study presented here. The idea of interconnection is illustrated in Fig. 1.

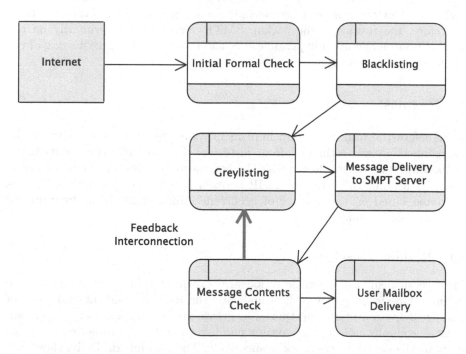

Fig. 1. Multi–layer spam detection system layout with feedback interconnection between Message contents check and Greylisting modules.

3.1   Interconnection Design Details

As explained above, the main aim of the interconnection being designed was to check whether the IP address of the SNMP sender of a message classified as spam in the Message contents check module is listed in Greylisting's AWL. If the IP address is listed, then the Greylisting module is instructed to remove it from AWL. For practical reasons, this function was implemented as:

1. a function collecting IP addresses of sending SMTP servers for every message classified as spam in Message contents check module, and
2. a batch file periodically (e.g., once a day) removing all IP addresses in the above list from Greylisting's AWL (hereinafter "AWL Cleaning").

3.2   Interconnection Testing

While the IP address collecting function (Item No. 1 in the numbered list in Sect. 3.1) was running permanently, the batch file (Item No. 2) was executed periodically once a month in the initial phase; later, the period was shortened to 1 day. The feedback interconnection was tested first on the backup SMTP server of the author's university (that is used for message reception only in periods of the primary SMTP server outages, i.e., quite seldom and for short periods). Then it was implemented to the primary SMTP server of the university. Therefore, the testing in the backup SMTP server focused primarily on the interconnection operation in periodic execution as well as integrity testing of the whole interconnection system.

4   Results

Even before the interconnection implementation, the expectable effect of the interconnection was estimated from historical data about incoming messages to the SMTP server for 30 months. The estimation results are shown in Fig. 2. The potential for the listing of an IP address into the Greylisting's AWL was evaluated based on the number of occurrences of a single IP address among message SNMP senders.

4.1   Results Overview

The overall results of spam detection are shown in Table 1. In the table, the columns Greylisted and Greyl. % indicate the monthly amount and ratio of message delivery attempts postponed by greylisting while Spam Cont. and Spam Cont. % columns indicate the monthly number and ratio of messages detected as spam in Message contents check, respectively, The column All. Deliv. shows the monthly number of messages allowed to be delivered, and the Total column shows the monthly number of attempts to deliver a message. Nevertheless, the results in this table cannot demonstrate the effect of the interconnection adequately.

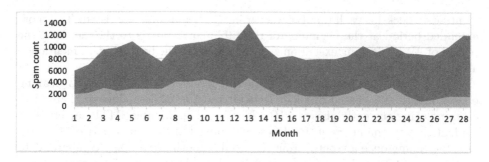

Fig. 2. Estimation of the interconnection effect on historical dates (horizontal axis - months, vertical axis - thousands of messages). The whole dark part (Blue + Red together) indicates the total number of spam messages detected by the Message contents check. The lower area (Red) means the portion of spam messages whose sender could potentially be listed to the Greylisting's AWL. (Color figure online)

Table 1. Results of spam detection in the implementation period

Month	Greylisted	Greyl. %	Spam Cont	Spam Cont. %	All. Deliv	Total attempts
1	75,991	12%	6,477	1%	114,215	635,788
2	74,288	13%	7,453	1.3%	115,216	569,968
3	106,409	4.2%	6,843	0.3%	138,224	2,553,333
4	97,946	15.4%	7,167	1.1%	148,101	636,585
5	93,751	15.5%	7,247	1.2%	155,719	603,162
6	99,922	12.8%	7,048	0.9%	140,451	782,896
7	104,850	12.8%	7,440	0.9%	155,979	821,590
8	104,035	11.2%	8,201	0.9%	151,895	932,641
9	109,087	11.6%	8,765	0.9%	150,642	943,559
10	117,064	9.7%	9,832	0.8%	152,273	1,211,425
11	121,786	14.5%	12,335	1.5%	148,838	842,797
12	122,243	16.9%	11,105	1.5%	134,531	721,417
13	156,839	22.4%	12,668	1.8%	123,386	699,539
14	115,415	25.3%	12,149	2.7%	123,945	455,777
15	123,513	20.1%	13,466	2.2%	150,516	615,638

4.2 Discussion on Results

As one can see in Table 1, the data shown here cannot verify the interconnection efficiency increase because the "natural" fluctuation in spam sources and thus the numbers of detected spam messages in individual months overlay the expected decrease in the "Spam Cont. %" column. For the sake of getting a better insight into obtained results, the more detailed data were tabulated in Table 2. The table shows the numbers of IP addresses removed from the Greylisting's AWL caused by each execution of the batch file described in Item 2 in Sect. 3.1. On the other hand, Table 3, where numbers of spam messages detected by the Message

Contents Check for each day (or month in the beginning) are listed. The more extended period at the beginning of the table corresponds to the fact that the period of AWL Cleaning execution was one month in the beginning, as mentioned above in Sect. 3.2.

When comparing both parts of Table 2, it looks clear that the AWL cleaning once a month (i.e., based on older IP addresses) is far less efficient than AWL cleaning using currently observed IP addresses of spamming SMTP servers. It is evident that the numbers of IP addresses removed in the upper part of the table (AWL cleaning once a month) from the whole month are roughly comparable to the number of IP addresses removed daily in the lower part of the table. Thus, the summary number of IP addresses removed in a month is many times lower in the upper part of the table. Therefore, despite the limited possibility of direct verification of the interconnection efficiency, a conclusion on the efficiency of the interconnection between Message Contents Check and Greylisting.

4.3   Statistical Evaluation

One of the critical issues to evaluate is always whether the obtained results are reliable in the sense of their repeatability. For getting at least a partial answer to this issue, statistical testing was performed. In spite of the fact that the massive fluctuations in daily spam counts due to the always–changing nature and frequency of spam messages, the comparison of average values listed in Table 3 gives a definite conclusion. The conclusion is that before AWL cleaning was implemented (the upper part of the table, the daily average is 1442 while in the period of monthly cleaning, the average value dropped to 575.5, and for daily cleaning, it further dropped to the value 433.6.

For a more detailed view, three statistical (null) hypotheses were formulated. The first hypothesis told that there is no statistically significant difference between periods with No AWL cleaning and Monthly cleaning (H1). The second hypothesis said that there was no statistically significant difference between the periods of Monthly AWL cleaning and Daily AWL cleaning (H2). In contrast, the third hypothesis told that there is no statistically significant difference between the periods of No AWL cleaning and Daily AWL cleaning (H3). After statistical calculations performed in JASP [14] (Student's t-test was used), the results, as summarized in Table 4, let us refuse all the three hypotheses. It is not surprising that the most significant difference was found in H3 (No cleaning vs. Daily cleaning). Nevertheless, the results in Table 4 mean that the statistically significant

Table 2. Numbers of IP addresses removed from Greylisting's AWL. The upper part shows the data for the period when the AWL Cleaning was executed once a month. The lower part of the table (starting Month 19, Day 16) indicates the daily numbers after daily AW cleaning was started.

Period of AWL cleaning		No. of IP addresses removed
Month	Day	
Month 2		3
Month 3		35
Month 4		34
Month 5		30
Month 6		32
Month 7		38
Month 13		19
Month 18		10
Month 19	Day 16	49
Month 19	Day 17	20
Month 19	Day 18	11
Month 19	Day 19	17
Month 19	Day 20	42
Month 19	Day 21	47
Month 19	Day 22	22
Month 19	Day 23	31
Month 19	Day 24	33
Month 19	Day 25	9
Month 19	Day 26	14
Month 19	Day 27	17
Month 19	Day 28	15
Month 19	Day 29	26
Month 19	Day 30	14
Month 19	Day 31	17
Month 20	Day 1	14
Month 20	Day 2	11
Month 20	Day 3	17

improvement of the whole multi-layer spam detection system (in the sense of a substantial decrease of spam messages detected in period with AWL cleaning reflecting better spam detection sensitivity of Greylisting) is proven after the application of the feedback connection described in Sect. 3.1.

Table 3. Daily numbers of spam messages (Spam column) detected in the Message Contents Check in the period of the implementation of the interconnection between Message Contents Check and Greylisting. The last column AWL cleaning indicate the period of AWL Cleaning.

Month	Day of the month	Spam count	AWL clearing
16	2	1275	Not used at all
16	3	1820	
16	4	1231	
17	1	925	Monthly
17	2	451	
17	3	1147	
17	4	585	
17	5	485	
17	6	216	
17	7	450	
17	8	758	
17	9	443	
17	10	295	
18	17	434	Daily
18	18	628	
18	19	704	
18	20	698	
18	21	629	
18	22	544	
18	23	269	
18	24	206	
18	25	243	
18	26	459	
18	27	452	
18	28	370	
18	29	538	
18	30	333	
18	31	318	
19	1	226	
19	2	452	
19	3	455	
19	4	647	
19	5	464	
19	6	374	
19	7	300	
19	8	256	

Table 4. Results of testing the 3 statistical hypotheses comparing periods of various AWL Cleaning

Hypothesis	t-statistics	Deg. of Freedom	p value
No vs. monthly (H1)	−4.45	11	<0.001
Monthly vs. daily (H2)	−1.832	31	0.077
No vs. daily (H3)	−9.317	24	<0.001

5   Conclusions

The above–described results demonstrate that the feedback interconnection between Message contents Check and Greylisting modules in a multi–level spam detecting system is significantly beneficial, yielding the significant increase of spam messages detected even before deliver (here in Greylisting). On the other hand, the positive effect is not extremely strong and can be easily overlaid by fluctuations in spam production. Nevertheless, the part of system resources necessary for reception and storage of spam messages that were not necessary to be stored were saved. Regarding the fact that the application of the interconnection technique described here is not limited to any specific implementation of the Greylisting[1] and Message Contents Check modules, it can be easily used in all multi–level spam detection systems with separated modules.

5.1   Future Research

Regarding the fact that both spam contents and techniques of its dissemination are continuously developed, it is necessary to keep up-to-date with the efficiency of tools and techniques applied in practical implementations, and, in the case of necessity, to adapt them to the changing "spam environment."

The interconnection between separate components on a multi–level spam detection system, as described here, does not represent the only option. Other components, like the Blacklisting module, can also be interconnected with other modules with potentially positive effects. This field has not been surveyed yet, but it deserves a reasonable research interest. The main result presented here, namely spam detection improvement through the feedback interconnection between Message-contents check and Greylisting, cannot be generalized directly into every spam detection system. Anyway, the results presented here still bring the way towards significant efficiency increase in many spam–detection instances.

Another factor with potentially high impact on spam detection and elimination is also worthy of mentioning, namely IPv6 transition. Despite the fact that application–layer services like e-mail are not directly influenced by the L3 protocols (either IP version 4 or IP version 6), the ongoing transition from the

---

[1] The only limitation of the interconnection described here consists in the fact that the existence of Greylisting's AWL is substantial. However, this is true for almost all Greylisting implementations.

legacy IP version 4 to the modern IP version 6 (IPv6) poses new challenges to spam detection mechanisms. The interconnection described in this paper is not directly related to any IP specific version but involved mechanisms (e.g., Greylisting implementation) work with IP addresses (so far mostly only with IPv4) so one can expect that the transition to a new implementation supporting IPv6 can change the behavior of the specific module. Therefore, this topic is also worth to attract some research interest.

References

1. Email & Spam data. Talos Intelligence. https://www.talosintelligence.com
2. Mehta, B., Hofmann, T.A.: Survey of attack-resistant collaborative filtering algorithms. IEEE Data Eng. Bull. **31**(2), 14–22 (2008)
3. Levine J.: DNS blacklists and whitelists. IETF RFC 5782 (2010). https://tools.ietf.org/html/rfc5782
4. Harris, E.: The next step in the spam control war: greylisting (2003). http://www.projects.puremagic.com/greylisting/whitepaper.html
5. Krause, T., Uetz, R., Kretschmann, T.: Recognizing email spam from meta data only. In: IEEE Conference on Communications and Network Security, pp. 178–186. IEEE (2019)
6. Habib, M., Faris, H., Hassonah, M.A., Alqatawna, J., Sheta, A.F., Al-Zoubi, A.M.: Automatic email spam detection using genetic programming with SMOTE. In: ITT 2018 - Information Technology Trends: Emerging Technologies for Artificial Intelligence, pp. 185–190 (2019)
7. Katasev, A.S., Emaletdinova, L.Y., Kataseva, D.V.: Neural network spam filtering technology. In: 2018 International Conference on Industrial Engineering, Applications and Manufacturing (2018)
8. Sochor, T.: Overview of e-mail SPAM elimination and its efficiency. In: IEEE 8th International Conference on Research Challenges in Information Science, pp. 191–201. IEEE (2014)
9. Lysenko, S., Savenko, O., Bobrovnikova, K., Kryshchuk, A.: Self-adaptive system for the corporate area network resilience in the presence of botnet cyberattacks. In: Gaj, P., Sawicki, M., Suchacka, G., Kwiecień, A. (eds.) Computer Networks 2018. CCIS, vol. 860, pp. 385–401. Springer, Heidelberg (2018). https://doi.org/10.1007/978-3-319-92459-5_31
10. Bartos, J., Walek, B., Klimes C., Farana R. Fuzzy application with expert system for conducting information security risk analysis. In: European Conference on Information Warfare and Security, ECCWS, pp. 33–41. University of Piraeus. Piraeus (2014)
11. Sochor, T., Davidova, A.: Potential of multilevel SPAM protection in the light of current SPAM trends. In: 10th IEEE International Conference on Networking, Sensing and Control. IEEE (2013)
12. Sochor, T., Farana, R.: Improving efficiency of e-mail communication via SPAM elimination using blacklisting. In: 21st Telecommunications Forum TELFOR School of Electrical Engineering, University of Belgrade (2013)
13. Tayal, D.K., Jain, A., Meena, K.: Development of anti-spam technique using modified K-Means & Naive Bayes algorithm. In: Proceeding of the 3rd International Conference on Computing for Sustainable Global Development, INDIACom (2016)
14. JASP Team. JASP (v 0.11.1) [Computer software] (2019)

Reliability Analysis of a Multipath Transport System in Fog Computing

Udo R. Krieger[1]($\boxtimes$) and Natalia M. Markovich[2]⊙

[1] Fakultät WIAI, Otto-Friedrich-Universität,
An der Weberei 5, 96047 Bamberg, Germany
udo.krieger@ieee.org
[2] V.A. Trapeznikov Institute of Control Sciences, Russian Academy of Sciences,
Profsoyuznaya Str. 65, Moscow 117997, Russia
markovic@ipu.rssi.ru

Abstract. We consider a fog computing approach with function virtualization in an IoT scenario that uses an SDN/NFV protocol stack and multipath communication between its clients and servers at the transport and session layers. We analyze the reliability of the associated redundant transport system comprising two logical channels that are susceptible to random failures. We model the error-prone system with a single repair unit and independent phase-type distributed repair times by a Marshall-Olkin failure model. The failure processes of both channels are described by general Markov-modulated Poisson processes (MMPPs) that are associated with the corresponding failure times and that are driven by the transitions of a common random environment. First we identify the generator matrix of the associated continuous-time Markov chain that is determined by the interarrival times of the Markov-modulated failure processes and the independent phase-type distributed repair times and the Kronecker-product structures of their associated parameter matrices. Then we show that the steady-state distribution of the restoration model can be effectively calculated by a semiconvergent iterative aggregation-disaggregation method for block matrices. Finally, we compute the associated reliability function and hazard rate of the multipath transport system.

Keywords: Fog computing · Marshall-Olkin failure model · Reliability function · Markov-modulated arrival process · Phase-type distributed repair times

1 Introduction

In recent years the cloud computing approach has been refined by mobile edge and fog computing to cover the technical challenges of new applications arising from the rapidly evolving Internet-of-Things (IoT) (cf. [1,2,8]). These architectures try to integrate new services based on advanced multimedia and machine-to-machine communication into the associated computing, storage, and internetworking infrastructures. They are based on modern software-defined networks

© Springer Nature Switzerland AG 2020
P. Gaj et al. (Eds.): CN 2020, CCIS 1231, pp. 101–116, 2020.
https://doi.org/10.1007/978-3-030-50719-0_9

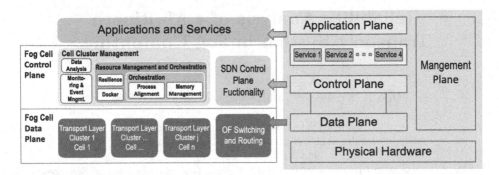

Fig. 1. Protocol stack of fog computing in a virtualized software environment.

(SDN), network function virtualization (NFV), and microservice concepts (cf. [14,24]). The fog computing architecture can be derived from the classical clear separation of functionalities into an application and services plane, a control plane, a data plane, and a management plane (see Fig. 1, cf. [7,9]).

In this context of client-server processing it has been realized that a multi-path communication which is established at the transport and session layers of the SDN/NFV protocol stack can substantially improve the capacity and reliability of the required fast interprocess communication. Adapted new transport layer protocols such as multipath TCP or multipath QUIC can be applied to establish redundant transport paths between clients and servers. The required multi-homing is realized by a use of multiple interfaces and will also be supported in the upcoming mobile settings of the local area and wide area 5G standards (see Fig. 2, cf. [3,5,20]).

In this paper we investigate a basic multipath transport system comprising two logical transport channels which provide redundant high-speed interprocess communication paths between a client as sender and a server as receiver in an SDN/NFV/5G-RAN environment. Due to the presumed existence of an exclusive, virtualized restoration function in the management plane, we describe the impact of a functional outage of each transport channel subject to random errors by a generalized Marshall-Olkin failure model (cf. [13,21–23]). We assume here that the entire redundant system is managed by a scalable, virtualized management system applying container virtualization techniques like Docker or Kubernetes (cf. [11,12]). It can instantiate a single repair function as virtual network function and provides a restoration of the original transport status after a generally distributed, nonnegative restoration period. We approximate the latter entity by random variables with phase-type (PH) distributions (cf. [25]).

Our main goal is to derive a Markovian reliability model of this redundant transport system in a random environment and to compute its reliability function and hazard rate. For this purpose we apply an effective computational solution method to a finite continuous-time Markov chain. In this way we enhance our related previous study [19] to a new setup of the underlying reliability model in a highly relevant technical context of fog computing. The new model exhibits a much more sophisticated algebraic structure of its generator matrix due to the involved three correlated Markov-modulated Poisson arrival processes of the failure patterns and the engagement of a single virtualized repairman function within the management layer of the fog computing architecture (see Fig. 3).

The paper is organized as follows. In Sect. 2 we describe the multipath transport system with two error-prone channels and a generalized Marshall-Olkin failure model. In Sect. 3 we derive the associated finite Markov chain with its generator matrix and calculate its related steady-state vector by an effective semiconvergent iteration scheme. In Sect. 4 we compute the reliability function and hazard rate of the multipath transport system. Finally, some conclusions and an outlook on further performance studies are presented.

2   Characterizing the Reliability of a Multipath Transport System by a Generalized Marshall-Olkin Model

We consider the hierarchical logical structure of a fog computing system that is deployed as edge computing infrastructure within the continuum between the cloud and the edge devices of an IoT environment (see Fig. 2, cf. [2,8]).

2.1   Description of a Transport System Related to Fog Computing

We assume that the fog computing environment is constructed by means of lightweight virtualization technology based on Linux containers and the orchestration framework Docker (cf. [9,12]). Dedicated stationary or mobile fog gateways provide the first logical entrance gates of its computing, storage and networking infrastructure. They realize aggregation points of the collected data streams generated by IoT sensor systems of corresponding smart edge devices in an associated fog cell (cf. Fig. 2, [9]). These generated data flows may be preprocessed and then forwarded to fog nodes in a higher logical layer or to processing and storage nodes in a distributed cloud infrastructure (cf. Fig. 2, see [7,26]). The latter nodes may also interact with a blockchain to support an immutable event history and the secure transfer of anonymous data elements in an underlying IoT infrastructure (cf. [7,26]). The minimal set of redundant transport paths between a fog node and more powerful nodes in the fog hierarchy or edge and cloud computing nodes, respectively, is modelled by two logical channels operating in hot stand-by mode. These entities are subject to random failures that may strike either one or both channels simultaneously. We suppose that each erroneous channel is immediately handled by independent repair activities which are triggered by a virtual surveillance function. The latter is realized in the management layer of the fog computing architecture (see Fig. 3).

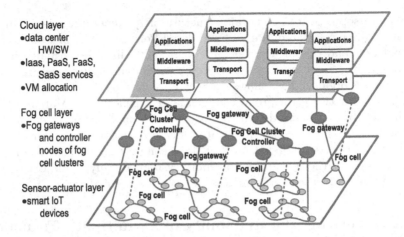

Fig. 2. A hierarchical fog computing architecture supporting IoT data processing in the fog cells (see also [7, 26]).

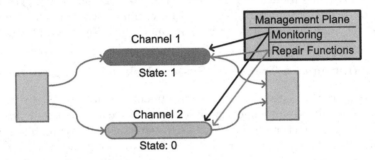

Fig. 3. A logical channel model of the redundant transport system with an operational channel 2 and an erroneous channel 1.

2.2 A Generalized Marshall-Olkin Failure Model of the Multipath Transport System

The considered multipath transport system comprises these two coupled logical channels. We suppose that they exhibit an identical logical structure. Therefore, we conclude that the error events are governed by a failure model of Marshall-Olkin type (see Figs. 2, 3, also [13, 21–23]).

The transfer function of each channel is hit by different kinds of errors that are triggered by a common internal or external environment. We can assume that a system failure occurs if the throughput along a channel drops below a certain predefined threshold or a total outage of the transport functions is observed. Then the latter error patterns are modelled by a Marshall-Olkin failure model with three correlated Markov-modulated Poisson processes (MMPPs) (cf. [10]) which are driving these failures on the individual channels 1 and 2, respectively, or strike both of them simultaneously (cf. [13, 21–23]). This MMPP class of Markovian arrival processes is an important subset the well-known general Markovian arrival processes (MAPs) (cf. [6]). We suppose that the latter MMPP processes can be described by the state of a common Markov-modulating environment $\{Y(t), t \geq 0\}$ in continuous time with a finite state space $\Sigma_Y = \{1, \ldots, K\}, K \in \mathbb{N}$ and an irreducible K−state generator matrix $Q \in \mathbb{R}^{K \times K}$. Its associated unique steady-state probability vector is denoted by $p \in \mathbb{R}^K$. It is determined by the solution of the linear system $p^T \cdot Q = 0, p^T \cdot e = 1$ with the vector of all ones $e \in \mathbb{R}^K$.

In the following, we apply the order relation $0 \ll x$ for vectors $x \in \mathbb{R}^N$. It shows that all components $x_i > 0$ of a vector $x \in \mathbb{R}^N$ are positive. In contrast, the order relation $0 < x$ indicates that $x \in \mathbb{R}^N$ is a nonnegative, non-zero vector, i.e. $0 \leq x_i$ for all $i \in \{1, \ldots N\}$ and $0 < x_i$ holds for at least one i (cf. [4]).

Considering a given state $Y(t) = j \in \Sigma_Y$ of the modulating environment, we assume that the interarrival times of any isolated failures imposed on channel 1 and 2 appear as independent exponentially distributed events with mean values $1/\lambda_{1j}$ and $1/\lambda_{2j}$, respectively, whereas a common failure is governed by the mean values $1/\lambda_{3j}$.

Let $0 \ll \lambda_1 = (\lambda_{11}, \ldots, \lambda_{1K})^T \in \mathbb{R}^K$, $0 \ll \lambda_2 = (\lambda_{21}, \ldots, \lambda_{2K})^T \in \mathbb{R}^K$, and $0 \ll \lambda_3 = (\lambda_{31}, \ldots, \lambda_{3K})^T \in \mathbb{R}^K$ be the positive column vectors of these associated arrival rates and $\Lambda_1 = \mathrm{Diag}(\lambda_1) > 0, \Lambda_2 = \mathrm{Diag}(\lambda_2) > 0, \Lambda_3 = \mathrm{Diag}(\lambda_3) > 0$ denote the corresponding diagonal-positive diagonal matrices of these arrival rate vectors of the failures in the random environment Y. Let $\Lambda = \Lambda_1 + \Lambda_2 + \Lambda_3$ be the arrival rate matrix of the superimposed MMPP arrival process of all correlated errors. Then the mean arrival rates of these three basic point processes are given by $\widehat{\lambda}_i = p^t \cdot \Lambda_i \cdot e = p^t \cdot \lambda_i, i \in \{1, 2, 3\}$, and $\widehat{\lambda} = p^t \cdot \Lambda \cdot e = \widehat{\lambda}_1 + \widehat{\lambda}_2 + \widehat{\lambda}_3$ holds (cf. [10]). We set $\overline{\Lambda}_1 = \Lambda_1 + \Lambda_3$ and $\overline{\Lambda}_2 = \Lambda_2 + \Lambda_3$ as arrival rate matrices of two corresponding MMPP processes that arise from a superposition of the streams 1 and 3 as well as 2 and 3, respectively.

Furthermore, we suppose that the initiated repair processes after an isolated error of channel 1 or 2, respectively, or the single maintenance process of both channels after a simultaneous outage are described by independent, phase-type distributed repair times R_1, R_2, R_3, respectively. Their stochastic characteristics are governed by general phase-type distributions

$$F_1(x) = \mathbb{P}\{R_1 \leq x\} = 1 - \beta^T \cdot \exp(T \cdot x) \cdot e, \tag{1}$$

$$F_2(x) = \mathbb{P}\{R_2 \leq x\} = 1 - \alpha^T \cdot \exp(S \cdot x) \cdot e, \tag{2}$$

$$F_3(x) = \mathbb{P}\{R_3 \leq x\} = 1 - \gamma^T \cdot \exp(U \cdot x) \cdot e \tag{3}$$

with the corresponding probability densities on the support set $[0, \infty) \subset \mathbb{R}$

$$f_1(x) = d\mathbb{P}\{R_1 \leq x\}/dt = \beta^T \cdot \exp(T \cdot x) \cdot T^0, \tag{4}$$

$$f_2(x) = d\mathbb{P}\{R_2 \leq x\}/dt = \alpha^T \cdot \exp(S \cdot x) \cdot S^0, \tag{5}$$

$$f_3(x) = d\mathbb{P}\{R_3 \leq x\}/dt = \gamma^T \cdot \exp(U \cdot x) \cdot U^0. \tag{6}$$

Here e denotes the vector of all ones of corresponding dimension. It means that three finite state phase-type representation matrices

$$(T, \beta), T \in \mathbb{R}^{n_1 \times n_1}, 0 < \beta \in \mathbb{R}^{n_1}, T^0 = -T \cdot e > 0, \tag{7}$$
$$(S, \alpha), S \in \mathbb{R}^{n_2 \times n_2}, 0 < \alpha \in \mathbb{R}^{n_2}, S^0 = -S \cdot e > 0, \tag{8}$$
$$(U, \gamma), U \in \mathbb{R}^{n_3 \times n_3}, 0 < \gamma \in \mathbb{R}^{n_3}, U^0 = -U \cdot e > 0 \tag{9}$$

with n_1, n_2, and n_3 states are used. Then the associated mean repair times are given by

$$\mathbb{E}(R_1) = 1/\mu_1 = -\beta^T \cdot T^{-1} \cdot e, \quad \mathbb{E}(R_2) = 1/\mu_2 = -\alpha^T \cdot S^{-1} \cdot e, \tag{10}$$
$$\mathbb{E}(R_3) = 1/\mu_3 = -\gamma^T \cdot U^{-1} \cdot e \tag{11}$$

and their variances are determined by

$$Var(R_1) = 2\beta^T \cdot T^{-2} \cdot e - (\beta^T \cdot T^{-1} \cdot e)^2, \tag{12}$$
$$Var(R_2) = 2\alpha^T \cdot S^{-2} \cdot e - (\alpha^T \cdot S^{-1} \cdot e)^2, \tag{13}$$
$$Var(R_3) = 2\gamma^T \cdot U^{-2} \cdot e - (\gamma^T \cdot U^{-1} \cdot e)^2. \tag{14}$$

Then the overall state of the multipath transport system can be described for $t \geq 0$ by a vector process

$$Z(t) = (X(t), M(t), Y(t)) = ((X_1(t), X_2(t)), (M_1(t), M_2(t), M_3(t)), Y(t)) \tag{15}$$

on the finite state space $\Sigma \subset \{0,1\}^2 \times \{0,1,\ldots,n_1\} \times \{0,1,\ldots,n_2\} \times \{0,1,\ldots,n_3\} \times \{1,\ldots,K\}$. The binary tuple $X(t) = (X_1(t), X_2(t)) = (i_1, i_2) \in \Sigma_X = \{0,1\}^2$ indicates by $X_1(t) = i_1 = 1$ or $X_2(t) = i_2 = 1$ that at time t a failure has occurred in channel 1 or 2, respectively, and the related channel is repaired by a virtual maintenance function of the transport system. A state $i_1 = 0$ or $i_2 = 0$ indicates a proper operation of the respective transport channel. $X(t) = (0,0)$ corresponds to the initial operational state and the failure state is determined by $X(t) = (1,1)$ where no further operation is possible until the maintenance process has been successfully executed on both channels. The common maintenance component

$$M(t) = (M_1(t), M_2(t), M_3(t)) = m = (m_1, m_2, m_3) \in \Sigma_M, \tag{16}$$

$\Sigma_M \subseteq \{0,1,\ldots,n_1\} \times \{0,1,\ldots,n_2\} \times \{0,1,\ldots,n_3\}$, records the phases $m = (m_1, m_2, m_3) \geq 0$ of the running repair processes for a state $i_1 = 1$ or $i_2 = 1$. Here a state $m_k = 0, k \in \{1,2,3\}$ indicates an idle repair function for a given related level component $i_l = 0, l \in \{1,2\}$.

We arrange the state variable $Z(t)$ and its overall state space Σ in such a way that the level process $X(t) = (X_1(t), X_2(t)) \in \Sigma_X$ is the leading indicator variable of the continuous-time Markov chain with the subspace $\Sigma_X = \{0,1,2,3\}$. Its four states are arranged according to a lexicographical ordering, i.e., it is

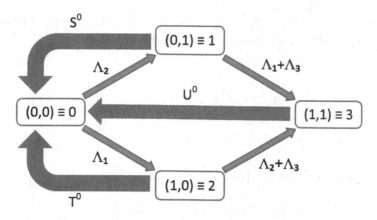

Fig. 4. Model of the states and transitions with their associated rate vectors and matrices of the related failure arrival epochs and maintenance completion events.

given by a binary encoding $0 \equiv (0,0), 1 \equiv (0,1), 2 \equiv (1,0), 3 \equiv (1,1)$. The phase variable $(M(t), Y(t)) \in \Sigma_{(M,Y)}$ with the phase state space

$$
\begin{aligned}
\Sigma_{(M,Y)} = \quad & \{(0,0,0)\} \times \{1,\ldots,K\} \\
\cup \ & \{0\} \times \{1,\ldots,n_2\} \times \{0\} \times \{1,\ldots,K\} \\
\cup \ & \{1,\ldots,n_1\} \times \{0\} \times \{0\} \times \{1,\ldots,K\} \\
\cup \ & \{0\} \times \{0\} \times \{1,\ldots,n_3\} \times \{1,\ldots,K\}
\end{aligned}
\tag{17}
$$

indicates the residual set of the microstates.

The initial state $Z(t) = z = (x, m, y)$ with $x = (0,0) \in \Sigma_X$ consists of the $j_0 = n_0 \cdot K = K, n_0 = 1$, microstates $(x, m, y) \in \{((0,0),(0,0,0))\} \times \{1,\ldots,K\}$, whereas the final error state with $x = (1,1) \in \Sigma_X$ comprises the $j_3 = n_3 \cdot K$ microstates $\{(1,1)\} \times \{(0,0)\} \times \{1,\ldots,n_3\} \times \{1,\ldots,K\}$. The two failure states with $x \in \{(0,1),(1,0)\} \subset \Sigma_X$ with one channel under repair consist of $j_1 = n_2 \cdot K$ and $j_2 = n_1 \cdot K$ microstates $\{(0,1)\} \times \{0\} \times \{1,\ldots,n_2\} \times \{0\} \times \{1,\ldots,K\}$ and $\{(1,0)\} \times \{1,\ldots,n_1\} \times \{0\} \times \{0\} \times \{1,\ldots,K\}$, respectively. A model of the state space with the associated transition vectors and matrices of the corresponding failure arrival epochs and maintenance completion events is illustrated in Fig. 4.

2.3 Analysis of a Simplified Marshall-Olkin Failure Model of the Redundant Transport System

We now consider the simplified Marshall-Olkin failure model with three independent Poisson input streams as failure triggers whose arrival rates are given by $\widehat{\lambda}_i = p^t \cdot \Lambda_i \cdot e = p^t \cdot \lambda_i, i \in \{1, 2, 3\}$. Furthermore, we may select the previously specified three independent phase-type distributed repair processes R_1, R_2, R_3 with the finite means $1/\mu_i = \mathbb{E}(R_i), i \in \{1, 2, 3\}$. Applying the steady-state results of Rykov et al. [23, Theorem 2], one can determine the steady-state probabilities $\Pi^{(S)} = (\Pi_0^{(S)}, \Pi_1^{(S)}, \Pi_2^{(S)}, \Pi_3^{(S)})^T \in \mathbb{R}^4$ of this simplified Marshall-Olkin failure model on the corresponding levels $i \in \Sigma_X = \{0,1,2,3\}$ in the following form:

$$\Pi_1^{(S)} = \Pi_0^{(S)} \cdot \frac{\widehat{\lambda}_1}{\widehat{\lambda}_2 + \widehat{\lambda}_3} \cdot \left[1 - \beta^T \cdot \left((\widehat{\lambda}_2 + \widehat{\lambda}_3)I - T \right)^{-1} \cdot T^0 \right] \tag{18}$$

$$\Pi_2^{(S)} = \Pi_0^{(S)} \cdot \frac{\widehat{\lambda}_2}{\widehat{\lambda}_1 + \widehat{\lambda}_3} \cdot \left[1 - \alpha^T \cdot \left((\widehat{\lambda}_1 + \widehat{\lambda}_3)I - S \right)^{-1} \cdot S^0 \right] \tag{19}$$

$$\Pi_3^{(S)} = \frac{\Pi_0^{(S)}}{\mu_3} \cdot \left(\widehat{\lambda}_1 \cdot \left[1 - \beta^T \cdot \left((\widehat{\lambda}_2 + \widehat{\lambda}_3)I - T \right)^{-1} \cdot T^0 \right] \right. \tag{20}$$
$$+ \widehat{\lambda}_2 \cdot \left[1 - \alpha^T \cdot \left((\widehat{\lambda}_1 + \widehat{\lambda}_3)I - S \right)^{-1} \cdot S^0 \right] + \widehat{\lambda}_3 \bigg)$$

$$\Pi_0^{(S)} = \left(1 + \frac{\widehat{\lambda}_1}{\widehat{\lambda}_2 + \widehat{\lambda}_3} \cdot \left[1 - \beta^T \cdot \left((\widehat{\lambda}_2 + \widehat{\lambda}_3)I - T \right)^{-1} \cdot T^0 \right] \cdot \left[1 + \frac{\widehat{\lambda}_2 + \widehat{\lambda}_3}{\mu_3} \right] \right. \tag{21}$$
$$\left. + \frac{\widehat{\lambda}_2}{\widehat{\lambda}_1 + \widehat{\lambda}_3} \cdot \left[1 - \alpha^T \cdot \left((\widehat{\lambda}_1 + \widehat{\lambda}_3)I - S \right)^{-1} \cdot S^0 \right] \cdot \left[1 + \frac{\widehat{\lambda}_1 + \widehat{\lambda}_3}{\mu_3} \right] + \frac{\widehat{\lambda}_3}{\mu_3} \right)^{-1}$$

This vector $\Pi^{(S)} \in \mathbb{R}^4$ can be used to approximate the initial steady-state solution $\alpha(x^{(0)})$ of the first aggregation system (47) that is triggering the disaggregation step (48) and the following iteration step (49) of the IAD approach in Subsect. 3.3.

3   Analyzing the Markov Model of the Transport System

In the following we investigate the finite continuous-time Markov chain (CTMC) $\{Z(t), t \geq 0\}$ that is used to analyze the reliability behavior of the described multipath transport system subject to the sketched generalized Marshall-Olkin failure model of its basic redundant, erroneous transport channels.

3.1   Generator Matrix of the Finite Markov Chain

In the following we consider the three different state sets $\{IOS, FS, IES\} \subset \mathcal{P}(\Sigma_X)$ comprising the initial operational state (IOS) $X(t) = x = (0,0) \in \Sigma_X$, the complete failure state (FS) $X(t) = x = (1,1) \in \Sigma_X$, and the cluster of isolated error states (IES) $X(t) = x \in \{(0,1),(1,0)\} \subset \Sigma_X$. Then the resulting generator matrix A of this finite CTMC $Z(t)$ has a block structure on the corresponding microstates $(x, m, y) \in \Sigma$ which is related to a redundant system with the Marshall-Olkin failure behavior (cf. [13, 21–23]):

$$A = \begin{pmatrix} A_{00} & A_{01} & A_{02} & A_{03} \\ A_{10} & A_{11} & 0 & A_{13} \\ A_{20} & 0 & A_{22} & A_{23} \\ A_{30} & 0 & 0 & A_{33} \end{pmatrix} \in \mathbb{R}^{N \times N}, \tag{22}$$

The transition behavior of the failure interarrivals in the random environment Y is driven by the irreducible generator matrix Q. The three PH-type driven repair processes are governed by (T, β), (S, α), (U, γ) that run independently of each other. Subsequently, we define the corresponding blocks A_{ij} of the generator matrix A in terms of the Kronecker product and Kronecker sum, i.e. $F \otimes E = (F_{ij} \cdot E)_{ij}$, and $F \oplus E = F \otimes I_l + I_m \otimes E$ for block matrices $F \in \mathbb{R}^{m \times m}$, $E \in \mathbb{R}^{l \times l}$ and identity matrices I_l, I_m as well as vectors of all ones e_l, e_m of appropriate dimensions $l > 0, m > 0$:

$$A_{00} = 1 \otimes (Q - \Lambda) = Q - \Lambda \tag{23}$$

$$A_{01} = 1 \otimes \alpha^T \otimes 1 \otimes \Lambda_2 = \alpha^T \otimes \Lambda_2 \tag{24}$$

$$A_{02} = \beta^T \otimes 1 \otimes 1 \otimes \Lambda_1 = \beta^T \otimes \Lambda_1 \tag{25}$$

$$A_{03} = 1 \otimes 1 \otimes \gamma^T \otimes \Lambda_3 = \gamma^T \otimes \Lambda_3 \tag{26}$$

$$A_{10} = 1 \otimes S^0 \otimes 1 \otimes I_K = S^0 \otimes I_K \tag{27}$$

$$A_{11} = 1 \otimes S \otimes 1 \otimes I_K + 1 \otimes I_{n_2} \otimes 1 \otimes (Q - \Lambda_1 - \Lambda_3)$$
$$= S \oplus (Q - \Lambda_1 - \Lambda_3) \tag{28}$$

$$A_{13} = 1 \otimes e_{n_2} \otimes \gamma^T \otimes (\Lambda_1 + \Lambda_3) = e_{n_2} \otimes \gamma^T \otimes (\Lambda_1 + \Lambda_3) \tag{29}$$

$$A_{20} = T^0 \otimes 1 \otimes 1 \otimes I_K = T^0 \otimes I_K \tag{30}$$

$$A_{22} = T \otimes 1 \otimes 1 \otimes I_K + I_{n_1} \otimes 1 \otimes 1 \otimes (Q - \Lambda_2 - \Lambda_3)$$
$$= T \oplus (Q - \Lambda_2 - \Lambda_3) \tag{31}$$

$$A_{23} = e_{n_1} \otimes 1 \otimes \gamma^T \otimes (\Lambda_2 + \Lambda_3) = e_{n_1} \otimes \gamma^T \otimes (\Lambda_2 + \Lambda_3) \tag{32}$$

$$A_{30} = 1 \otimes 1 \otimes U^0 \otimes I_K = U^0 \otimes I_K \tag{33}$$

$$A_{33} = 1 \otimes 1 \otimes U \otimes I_K + 1 \otimes 1 \otimes I_{n_3} \otimes Q = U \oplus Q \tag{34}$$

$$A_{31} = A_{32} = A_{12} = A_{21} = 0 \tag{35}$$

Then the part of the generator matrix A on the operational states $OS = \{0, 1, 2\} \equiv \{(0,0), (0,1), (1,0)\} \subset \Sigma_X$ excluding the failure state $FS = \{3\} \equiv \{(1,1)\} \subset \Sigma_X$ is defined by the block matrix

$$A_O = \begin{pmatrix} A_{00} & A_{01} & A_{02} \\ A_{10} & A_{11} & 0 \\ A_{20} & 0 & A_{22} \end{pmatrix} \in \mathbb{R}^{M \times M} \tag{36}$$

with $M = K \cdot (n_0 + n_1 + n_2) = K \cdot (1 + n_1 + n_2)$ states.

3.2   Calculation of the Steady-State Vector

In the following we suppose that an irreducible generator matrix Q of the Markovian environment and three irreducible phase-type generators $T + T^0 \beta^T$, $S + S^0 \alpha^T$, $U + U^0 \gamma^T$ are given. Then we denote by $\Pi^T = (\Pi_0^T, \Pi_1^T, \Pi_2^T, \Pi_3^T) \gg 0$ the resulting partitioned, unique steady-state row vector of the irreducible Markov chain $Z(t)$.

We can calculate Π by efficient numerical solution methods for finite ergodic Markov chains such as direct or iterative solution techniques of the balance equations $\Pi^T \cdot A = 0$, $\Pi^T \cdot e = 1$, for instance, by applying aggregation-disaggregation methods such as an additive or multiplicative Schwarz decomposition method or any other iteration scheme derived from an M-splitting (cf. [4,15–18,25]).

Let $\tilde{A} = -A^T$ denote the irreducible M-matrix associated with the generator matrix A and $A = L + U - \Delta$ be the Jacobi block-matrix decomposition into the diagonal block matrix $\Delta = -\mathrm{Diag}(A_{00}, A_{11}, A_{22}, A_{33})$, and lower- and upper-diagonal block matrices

$$
L = \begin{pmatrix} 0 & 0 & 0 & 0 \\ A_{10} & 0 & 0 & 0 \\ A_{20} & 0 & 0 & 0 \\ A_{30} & 0 & 0 & 0 \end{pmatrix}, \quad
U = \begin{pmatrix} 0 & A_{01} & A_{02} & A_{03} \\ 0 & 0 & 0 & A_{13} \\ 0 & 0 & 0 & A_{23} \\ 0 & 0 & 0 & 0 \end{pmatrix}, \tag{37}
$$

respectively. Then we define the associated M-splitting $\tilde{A} = -A^T = M - N$ with the corresponding transposed matrices of the block-matrix decomposition

$$
M = \Delta^T = - \begin{pmatrix} A_{00}{}^T & 0 & 0 & 0 \\ 0 & A_{11}{}^T & 0 & 0 \\ 0 & 0 & A_{22}{}^T & 0 \\ 0 & 0 & 0 & A_{33}{}^T \end{pmatrix} \tag{38}
$$

$$
N = L^T + U^T = \begin{pmatrix} 0 & A_{10}{}^T & A_{20}{}^T & A_{30}{}^T \\ A_{01}{}^T & 0 & 0 & 0 \\ A_{02}{}^T & 0 & 0 & 0 \\ A_{03}{}^T & A_{13}{}^T & A_{23}{}^T & 0 \end{pmatrix}. \tag{39}
$$

We get the iteration matrix $J = M^{-1} \cdot N = [\Delta^T]^{-1} \cdot [L^T + U^T]$ and the associated nonnegative matrix

$$
\tilde{T} = I_N - \tilde{A} \cdot M^{-1} = N \cdot M^{-1} \tag{40}
$$

$$
= \begin{pmatrix} 0 & A_{10}{}^T \cdot [-A_{11}]^{-T} & A_{20}{}^T \cdot [-A_{22}]^{-T} & A_{30}{}^T \cdot [-A_{33}]^{-T} \\ A_{01}{}^T \cdot [-A_{00}]^{-T} & 0 & 0 & 0 \\ A_{02}{}^T \cdot [-A_{00}]^{-T} & 0 & 0 & 0 \\ A_{03}{}^T \cdot [-A_{00}]^{-T} & A_{13}{}^T \cdot [-A_{11}]^{-T} & A_{23}{}^T \cdot [-A_{22}]^{-T} & 0 \end{pmatrix}
$$

$$
\tilde{T} = \begin{pmatrix} 0 & (S^0)^T \otimes I_K & (T^0)^T \otimes I_K & (U^0)^T \otimes I_K \\ \alpha \otimes \Lambda_2 & 0 & 0 & 0 \\ \beta \otimes \Lambda_1 & 0 & 0 & 0 \\ \gamma \otimes \Lambda_3 & [e_{n_2}]^T \otimes \gamma \otimes (\Lambda_1 + \Lambda_3) & [e_{n_1}]^T \otimes \gamma \otimes (\Lambda_2 + \Lambda_3) & 0 \end{pmatrix} \tag{41}
$$
$$
\cdot \begin{pmatrix} [Q^T - \Lambda]^{-1} & 0 & 0 & 0 \\ 0 & [S^T \oplus (Q^T - \Lambda_1 - \Lambda_3)]^{-1} & 0 & 0 \\ 0 & 0 & [T^T \oplus (Q^T - \Lambda_2 - \Lambda_3)]^{-1} & 0 \\ 0 & 0 & 0 & [U^T \oplus Q^T]^{-1} \end{pmatrix}
$$

with the property $M^{-1} \cdot \tilde{T} \cdot M = J$. This stochastic matrix $\tilde{T}$ extends the structure of the Marshall-Olkin reliability model to a block matrix. Its reduction to single elements by means of the aggregation-disaggregation approach will yield a stochastic matrix $B \equiv B(x) = R \cdot T \cdot P(x) \in \mathbb{R}^{4 \times 4}$ subject to an aggregation matrix R and a prolongation matrix $P(x)$ for any probability vector $0 < x \in \mathbb{R}^N$, $e^T \cdot x = 1$ (see (43)). The latter reflects the connectivity graph of the original Marshall-Olkin reliability model.

Then the column-stochastic block structured matrix

$$T = I_N - \omega \widetilde{A} \cdot M^{-1} = (1 - \omega)I_N + \omega \widetilde{T} \qquad (42)$$

is a semiconvergent, nonnegative matrix for any scaling $\omega \in (0, 1)$ (cf. [4, 15, 25]). Thus, the algebraically similar iteration matrix $J_\omega = I_N - \omega M^{-1} \cdot \widetilde{A} = (1 - \omega)I_N + \omega \widetilde{J}$ which has the same spectrum as T is also semiconvergent for any $\omega \in (0, 1)$.

Based on the block-matrix decomposition of A in (22) we determine a partition $\Gamma = \{J_0, J_1, J_2, J_3\}$ into $m = 4$ subsets of the state space $\Sigma = \{1, \ldots, N\}, N = n_0 \cdot K + n_2 \cdot K + n_1 \cdot K + n_3 \cdot K, n_0 = 1$, with the four disjoint subsets $J_0 = \{1, \ldots, K\}, J_1 = \{K + 1, \ldots, (1 + n_2) \cdot K\}, J_2 = \{(1 + n_2) \cdot K + 1, \ldots, (1 + n_1 + n_2) \cdot K\}, J_3 = \{(1 + n_1 + n_2) \cdot K + 1, \ldots, (1 + n_1 + n_2 + n_3) \cdot K\}$.

3.3 Application of a Semiconvergent IAD-Algorithm for M-Matrices

In the following we apply an iterative aggregation-disaggregation (IAD) algorithm that is semiconvergent to the unique steady-state vector $\Pi^T = (\Pi_i^T)_i \gg 0$ with its positive components Π_i^T on the partition set J_i for each state $i \in \{0, 1, 2, 3\}$ (cf. [17,18]). The IAD-algorithm includes three basic matrices. First, we generate an aggregation matrix R and a prolongation matrix $P(x)$,

$$R = \begin{pmatrix} e_{J_0}^T & 0 & 0 & 0 \\ 0 & e_{J_1}^T & 0 & 0 \\ 0 & 0 & e_{J_2}^T & 0 \\ 0 & 0 & 0 & e_{J_3}^T \end{pmatrix} \in \mathbb{R}^{4 \times N}, \quad P(x) = \begin{pmatrix} y_0 & 0 & 0 & 0 \\ 0 & y_1 & 0 & 0 \\ 0 & 0 & y_2 & 0 \\ 0 & 0 & 0 & y_3 \end{pmatrix} \in \mathbb{R}^{N \times 4}, \quad (43)$$

for $0 < x = \begin{pmatrix} x_0 \\ x_1 \\ x_2 \\ x_3 \end{pmatrix} \in \mathbb{R}^N, e^T \cdot x = 1$, in terms of

$$[\alpha(x)]_j = e_{J_j}^T \cdot x_j \qquad [y(x)]_j = x_j / [\alpha(x)]_j \qquad (44)$$

provided that $x_j > 0$ holds for its component on set $J_j, j \in \{0, 1, 2, 3\}$, and we use the uniform distribution in case of $x_j = 0$ for a given j. Here $e_{J_0} \in \mathbb{R}^{n_0 K}, e_{J_1} \in \mathbb{R}^{n_2 K}, e_{J_2} \in \mathbb{R}^{n_1 K}, e_{J_3} \in \mathbb{R}^{n_3 K}, e_4 \in \mathbb{R}^4, e \in \mathbb{R}^N$ denote the vectors of all ones.

Applying the iteration matrix $T = T(\omega)$ in (42) and such a nonnegative vector $x \in \mathbb{R}^N$ we get the corresponding aggregated matrix $B(x) \in \mathbb{R}^{4 \times 4}$ in terms of

$$B(x) = R \cdot T \cdot P(x). \qquad (45)$$

We further use $r(x) = ||(I_N - T) \cdot x||_1$ with the identity matrix I_N for the L_1-norm $||x||_1 = \sum_1^N |x_i|$ in $\mathbb{R}^N$.

Then the IAD-algorithm reads as follows:

1. We choose four real numbers $\omega, \epsilon, c_1, c_2 \in (0, 1)$ and set $k = 0$.
   First, we select the steady-state vector of the simplified Marshall-Olkin model as initial vector $\alpha(x^{(0)}) = \Pi^{(S)} \in \mathbb{R}^4$, cf. (18)–(21). Then we construct the initial probability vector $x^{(0)} = (x_i^{(0)})_i \gg 0, e^T \cdot x^{(0)} = 1$, by expanding $\alpha(x^{(0)})$ uniformly on each subset $J_i, i \in \{0, 1, 2, 3\}$ in terms of

$$x_i^{(0)} = \frac{\alpha(x^{(0)})_i}{n_i} \cdot e_{J_i} \tag{46}$$

Then we go to step 3.
2. We solve

$$B(x^{(k)}) \cdot \alpha(x^{(k)}) = \alpha(x^{(k)}) \tag{47}$$

subject to $\quad e_4^T \cdot \alpha(x^{(k)}) = 1, \quad \alpha(x^{(k)}) > 0.$
3. We compute

$$\tilde{x} = P(x^{(k)}) \cdot \alpha(x^{(k)}). \tag{48}$$

4. We compute

$$x^{(k+1)} = T \cdot \tilde{x}. \tag{49}$$

5. If

$$r(\tilde{x}) \le c_1 \cdot r(x^{(k)})$$

        then   go to step 6
    else   compute

$$x^{(k+1)} = T^h \cdot \tilde{x} \tag{50}$$

    for $h > 1$ such that $\quad r(x^{(k+1)}) \le c_2 \cdot r(x^{(k)})$
    endif
6. If

$$||x^{(k+1)} - x^{(k)}||_1 / ||x^{(k)}||_1 < \epsilon \tag{51}$$

       then   go to step 7
   else

$$k = k + 1,$$

       and   go to step 2
   endif
7. At the end we perform a normalization after a successful convergence test:

$$\Pi = \frac{M^{-1} \cdot x^{(k+1)}}{e^T \cdot M^{-1} \cdot x^{(k+1)}} \tag{52}$$

The existing convergence theory related to numerical solution methods for finite Markov chains has revealed that the semiconvergence of this specific IAD-algorithm to the probability vector Π can be proven (cf. [17,18,25]). The specific selection of its initial vector $x^{(0)}$ in (46) by the simplified Marshall-Olkin model shall guarantee this required local convergence behavior of our approach.

4   Computing the Reliability Function and Hazard Rate of the Redundant Transport System

The reliability of the error-prone multipath transport system is characterized by the dwell time $D_T \geq 0$ in the set of the operational states $\widehat{O} = \{z = (x, h, y) \in \Sigma \mid x \in OS \subset \Sigma_X\}$ of the overall state space Σ subject to the start in one of those states $z \in \widehat{O}$ in the steady-state regime with the steady-state row vector $\Pi_{\widehat{O}}^T = (\Pi_0^T, \Pi_1^T, \Pi_2^T) \gg 0$ and its positive components $\Pi_i^T \gg 0$ associated with each non-failure state $i \in OS = \{0, 1, 2\} \subset \Sigma_X$.

Then we can calculate the reliability function $F_R(t) = \mathbb{P}\{D_T > t\}$ as time-dependent probability of the Markov chain $Z(t)$ to reside in a state $z \in \widehat{O}$ up to time $t > 0$ given that a capturing in the absorbing states $\widehat{F} = \{z = (x, h, y) \in \Sigma \mid x \in FS \subset \Sigma_X\}$ does not occur before that epoch (cf. (36), see also [13,19]):

$$F_R(t) = \mathbb{P}\{Z(0) \in \widehat{O}\} \cdot \mathbb{P}\{D_T > t \mid Z(0) \in \widehat{O}\}$$
$$= \mathbb{P}\{Z(0) \in \widehat{O}\} \cdot \mathbb{P}\{Z(t) \notin \widehat{F} \mid Z(0) \in \widehat{O}\} = \Pi_{\widehat{O}}^T \cdot \exp(A_O t) \cdot e \quad (53)$$

The computation of the matrix exponential $\exp(A_O t)$ can be effectively performed by means of a uniformization approach (cf. [25]).

Let $D = \mathrm{Diag}(D_{ii}) > 0$ denote the diagonal matrix determined by the positive diagonal elements $D_{ii} = -(A_O)_{ii} > 0, i \in \{1, \ldots, M\}$, of the M-matrix $-A_O$ in (36). We define a constant $\gamma = \max_{1 \leq i \leq M}(D_{ii}) > 0$ and use the substochastic submatrix of the generator matrix $P_O = (P_{ij}), 1 \leq i \leq M, 1 \leq j \leq M$, that is determined by the transition probabilities $P_O = I_M + A_O/\gamma$ at the embedded time epochs of transition events in the Markov chain $Z(t)$. Then it holds $A_O = \gamma \cdot (P_O - I_M)$ with the identity matrix $I_M \in \mathbb{R}^{M \times M}$. It induces a simple representation as matrix exponential

$$R_O(t) = \exp(A_O \cdot t) = \exp(\gamma t \cdot (P_O - I_M))$$

$$= \left[\sum_{n=0}^{\infty} \frac{(\gamma t)^n}{n!} \exp(-\gamma t) \cdot ((P_O)^n)_{ij} \right]_{1 \leq i,j \leq M}. \quad (54)$$

The latter form allows a fast computation of the reliability function $F_R(t)$ in (53) in terms of the matrix-vector product

$$F_R(t) = \Pi_{\widehat{O}}^T \cdot R_O(t) \cdot e = \sum_{n=0}^{\infty} \frac{(\gamma t)^n}{n!} \exp(-\gamma t) \cdot \Pi_O^T \cdot (P_O)^n \cdot e \quad (55)$$

by means of a Poisson distribution with parameter γt and the consecutive powers of the sub-stochastic matrix P_O (cf. [19,25]).

If we assume to start in steady state with the probability distribution Π_O, then the hazard rate $h(t) : [0, \infty) \to \mathbb{R}$ of the reliability model can be simply calculated by the Markov chain with the absorbing failure states $z \in \widehat{F}$ of $FS = \{(1,1)\}$ as a simple phase-type model:

$$h(t) = \frac{d[1 - F_R(t)]}{dt} \cdot [F_R(t)]^{-1} = \frac{\Pi_O^T \cdot \exp(A_O t) \cdot (-A_O) \cdot e}{\Pi_O^T \cdot \exp(A_O t) \cdot e} \quad t \geq 0. \quad (56)$$

Due to $A \cdot e = 0$ and $\gamma^T \cdot e = 1$ we get

$$\Delta = -A_O \cdot e = \begin{pmatrix} A_{03} \\ A_{13} \\ A_{23} \end{pmatrix} \cdot e = \begin{pmatrix} \gamma^T \otimes \Lambda_3 \\ e_{n_2} \otimes \gamma^T \otimes (\Lambda_1 + \Lambda_3) \\ e_{n_1} \otimes \gamma^T \otimes (\Lambda_2 + \Lambda_3) \end{pmatrix} \cdot e$$

$$= \begin{pmatrix} \gamma^T \otimes \lambda_3 \\ e_{n_2} \otimes \gamma^T \otimes (\lambda_1 + \lambda_3) \\ e_{n_1} \otimes \gamma^T \otimes (\lambda_2 + \lambda_3) \end{pmatrix} \cdot e = \begin{pmatrix} \lambda_3 \\ e_{n_2} \otimes (\lambda_1 + \lambda_3) \\ e_{n_1} \otimes (\lambda_2 + \lambda_3) \end{pmatrix} \quad (57)$$

with vectors of all ones e of appropriate dimensions. Inserting the uniformization representation (55), we conclude that

$$h = \lim_{t \to \infty} h(t) = \frac{\Pi_O^T \cdot P_O \cdot (-A_O) \cdot e}{\Pi_O^T \cdot P_O \cdot e}$$

$$= \frac{\Pi_O^T \cdot P_O \cdot \Delta}{\Pi_O^T \cdot [e - \Delta/\gamma]} = \frac{\Pi_O^T \cdot P_O \cdot \Delta}{1 - \Pi_O^T \cdot \Delta/\gamma} \quad (58)$$

holds for the asymptotic regime $t \to \infty$ and we approach a corresponding exponential distribution with mean $1/h$ in this asymptotic regime. This outcome expands the results of Kozyrev, Rykov et al. [13,22] to the developed generalized Marshall-Olkin failure model.

5   Conclusions

We have considered the application of a fog computing paradigm with function virtualization to an IoT scenario that is supported by an SDN/NFV protocol stack and a multipath communication between its clients and servers (cf. [8,14, 20,24]).

It has been our major goal to model this error-prone multipath transport system with a single repairman and independent phase-type distributed repair times by a generalized Marshall-Olkin failure model. For this purpose the failure processes of the incorporated two logical transport channels between a client-server pair have been described by three Markov-modulated Poisson failure processes that are driven by the transitions of a common random environment. The restoration processes were modelled by general phase-type distributed repair times.

First we have identified the generator matrix of the derived finite, continuous-time Markov chain of the reliability model in terms of associated Kronecker products of the parameter matrices which are related to the Markov-modulated interarrival times of failures and the phase-type distributed repair times. Then we have revealed that the steady-state distribution of this restoration model of Marshall-Olkin type can be effectively computed by means of an iterative aggregation-disaggregation method that has been derived from a Jacobi splitting of an associated block structured M-matrix. The latter scheme has used a closed-form representation of the steady-state vector of a simpler Marshall-Olkin failure model derived by Rykov, Kozyrev et al. [23]. Finally, we have used this outcome to compute the reliability function and the hazard rate of the multipath transport system by means of an appropriately defined finite, absorbing Markov chain and we have revealed its form in the asymptotic regime of time.

Our future work will focus on the sensitivity analysis of the reliability function and hazard rate with regard to the properties of the Markov-modulated arrival processes. Moreover, we shall consider the application of the Marshall-Olkin failure model to other services in SDN/NFV networks with an integrated 5G RAN that can support fog and mobile edge computing (cf. [1,7,14]).

Acknowledgment. N.M. Markovich was partly supported by the Russian Foundation for Basic Research (grant 19-01-00090).

References

1. Aazam, M., Huh, E.-N.: Fog computing and smart gateway based communication for cloud of things. In: 2014 International Conference on Future Internet of Things and Cloud (FiCloud), Barcelona, Spain, 27–29 August, pp. 464–470 (2014)
2. Al-Fuqaha, A., et al.: Internet of Things: a survey on enabling technologies, protocols, and applications. IEEE Commun. Surv. Tutor. **17**(4), 2347–2376 (2015)
3. Barré, S., Paasch, C., Bonaventure, O.: MultiPath TCP: from theory to practice. In: Domingo-Pascual, J., Manzoni, P., Palazzo, S., Pont, A., Scoglio, C. (eds.) NETWORKING 2011. LNCS, vol. 6640, pp. 444–457. Springer, Heidelberg (2011). https://doi.org/10.1007/978-3-642-20757-0_35
4. Berman, A., Plemmons, R.J.: Nonnegative Matrices in the Mathematical Sciences. Academic Press, New York (1979)
5. Bonaventure, O., et al.: Multipath QUIC. In: CoNEXT 2017, Seoul/Incheon, South Korea, 12–15 December (2017). See also https://multipath-quic.org/
6. Chakravarthy, S. R.: Markovian Arrival Processes. In: Wiley Encyclopedia of Operations Research and Management Science (2011)
7. Cech, H., Großmann, M., Krieger, U.R.: A fog computing architecture to share sensor data by means of blockchain functionality. In: IEEE International Conference on Fog Computing, Prague, Czech Republic, 24–26 June (2019)
8. Chiang, M., Zhang, T.: Fog and IoT: an overview of research opportunities. IEEE Internet Things J. **3**(6), 854–864 (2016)
9. Eiermann, A., Renner, M., Großmann, M., Krieger, U.R.: On a fog computing platform built on ARM architectures by Docker container technology. In: Eichler, G., Erfurth, C., Fahrnberger, G. (eds.) I4CS 2017. CCIS, vol. 717, pp. 71–86. Springer, Cham (2017). https://doi.org/10.1007/978-3-319-60447-3_6

10. Fischer, W., Meier-Hellstern, K.: The Markov-modulated Poisson process (MMPP) cookbook. Perform. Eval. **18**(2), 149–171 (1993)
11. Großmann, M., Ioannidis, C.: Continuous integration of applications for ONOS. In: 5th IEEE Conference on Network Softwarization (NetSoft 2019), Paris, France, 24–28 June (2019)
12. Holla, S.: Orchestrating Docker. Packt Publishing Ltd., Birmingham (2015)
13. Kozyrev, D., Rykov, V., Kolev, N.: Reliability function of renewable system with Marshall-Olkin failure model. Reliab. Theory Appl. **13**(1(48)), 39–46 (2018)
14. Kozyrev, D., et al.: Mobility-centric analysis of communication offloading for heterogeneous Internet of Things devices. Wireless Commun. Mob. Comput. **2018** (2018). Article ID 3761075
15. Krieger, U.R., Müller-Clostermann, B., Sczittnick, M.: Modeling and analysis of communication systems based on computational methods for Markov chains. IEEE J. Sel. Areas Commun. **8**(9), 1630–1648 (1990)
16. Krieger, U.R.: Analysis of a loss system with mutual overflow in a Markovian environment. In: Stewart, W. (ed.) Numerical Solution of Markov Chains, pp. 303–328. Marcel Dekker, New York (1990)
17. Krieger, U.R.: On a two-level multigrid solution method for finite Markov chains. Linear Algebra Appl. **223**(224), 415–438 (1995)
18. Krieger, U.R.: Numerical solution of large finite Markov chains by algebraic multigrid techniques. In: Stewart, W. (ed.) Computations with Markov Chains, pp. 403–424. Kluwer Academic Publishers, Boston (1995). https://doi.org/10.1007/978-1-4615-2241-6_23
19. Krieger, U.R., Markovich, N.: Modeling and reliability analysis of a redundant transport system in a Markovian environment. In: Vishnevskiy, V.M., Samouylov, K.E., Kozyrev, D.V. (eds.) DCCN 2019. LNCS, vol. 11965, pp. 302–314. Springer, Cham (2019). https://doi.org/10.1007/978-3-030-36614-8_23
20. Rimal, B.P., Van, D.P., Maier, M.: Mobile edge computing empowered fiber-wireless access networks in the 5G era. IEEE Commun. Mag. **55**, 192–200 (2017)
21. Rykov, V.V., Kozyrev, D.V.: Analysis of renewable reliability systems by markovization method. In: Rykov, V.V., Singpurwalla, N.D., Zubkov, A.M. (eds.) ACMPT 2017. LNCS, vol. 10684, pp. 210–220. Springer, Cham (2017). https://doi.org/10.1007/978-3-319-71504-9_19
22. Rykov, V., Zaripova, E., Ivanova, N., Shorgin, S.: On sensitivity analysis of steady state probabilities of double redundant renewable system with Marshall-Olkin failure model. In: Vishnevskiy, V.M., Kozyrev, D.V. (eds.) DCCN 2018. CCIS, vol. 919, pp. 234–245. Springer, Cham (2018). https://doi.org/10.1007/978-3-319-99447-5_20
23. Rykov, V., Dimitrov, B.: Renewal redundant systems under the Marshall-Olkin failure model. Sensitivity analysis. In: Vishnevskiy, V.M., Samouylov, K.E., Kozyrev, D.V. (eds.) DCCN 2019. LNCS, vol. 11965, pp. 234–248. Springer, Cham (2019). https://doi.org/10.1007/978-3-030-36614-8_18
24. Stallings, W.: Foundations of Modern Networking: SDN, NFV, QoE, IoT, and Cloud. Pearson, Indianapolis (2016)
25. Stewart, W.J.: Probability, Markov Chains, Queues, and Simulation. Princeton University Press, Princeton (2009)
26. Ziegler, H.M., et al.: Integration of fog computing and blockchain technology using the plasma framework. In: Proceedings of the EEE International Conference on Blockchain and Cryptocurrency (ICBC 2019), May 2019, Seoul, South Korea (2019)

Minimising the Churn Out of the Service by Using a Fairness Mechanism

Izabela Mazur[✉], Jacek Rak, and Krzysztof Nowicki

Faculty of Electronics, Telecommunications and Informatics,
Gdansk University of Technology, Gdansk, Poland
izabela.mazur@onet.com.pl, {jrak,krzysztof.nowicki}@pg.edu.pl

Abstract. The paper proposes an algorithm of bandwidth distribution, ensuring fairness to end-users in computer networks. The proposed algorithm divides users into satisfied and unsatisfied users. It provides fairness in terms of quality of experience (QoE) for satisfied users and quality of service (QoS) for unsatisfied users. In this paper, we present detailed comparisons relevant to service providers to show the advantages of the proposed algorithm over the popular max-min algorithm. Our algorithm is designed to provide service providers with a mechanism to minimize the number of end-user terminations of service, which is one of the most desired factors for service providers.

Keywords: Fairness · QoE · QoS · Churn

1 Introduction

In recent years, network traffic associated with Internet multimedia services such as video streaming, online games and online stock exchanges has become the largest part of Internet traffic, and this phenomenon is expected to continue to grow [1]. This is a challenge to provide quality services to a large number of users who use them. It is a demanding and non-trivial challenge. Currently, one of the most commonly used methods of providing video services is the Dynamic Adaptive Streaming over HTTP (DASH) standard. It is based on dividing content into sequences of small HTTP-based files. Each of these files contains a short fragment of the transmission of playback content, no more than a few seconds. These files which are called segments are transmitted at different bit rates and later are combined into single consistent content. This is to minimize the number of interruptions in playback that can occur due to changing network conditions [2]. DASH standard is designed to ensure high level of usability and stable quality of service (QoS) and to increase the sense for quality of experience (QoE) which is extremely important in the context of maintaining the users with a given provider and minimizing the number of customers resignations from a given provider.

For teleinformation and telecommunication service providers, minimization of service resignation by users, i.e., the phenomenon of churn, is an extremely

© Springer Nature Switzerland AG 2020
P. Gaj et al. (Eds.): CN 2020, CCIS 1231, pp. 117–137, 2020.
https://doi.org/10.1007/978-3-030-50719-0_10

important challenge, because it is associated with the loss of profits and the need to incur additional costs of acquiring a new customer, which are significantly higher than the costs of maintaining the customer.

One of the factors influencing churn is the level of fairness in a given system. This is particularly visible in the case of online games, where the user can feel fair or unfair relating to other players. Another example is the access to multimedia content in the same subnet, e.g., in home subnet, where the user is aware of QoE of another users. Lack of fairness towards other users leads to frustration among users and abandonment of the service.

Concerning the characteristics of multimedia services, the quality of experience and the awareness of the need for a sense of fairness among users, in this paper we propose a mechanism ensuring a fair distribution of technical parameters in relation to QoE for users who will remain with the service provider and in relation to QoS for users who may give up the service. At the same time, minimum number of leaving users from a given provider is ensured.

1.1 Background

There are many definitions of fairness. Some of them: "Fairness is the quality of judgments free from discrimination" [3]. "Fair and equitable treatment or conduct without favoritism or discrimination" [4]. "Quality of treatment of people in equal measure or in a proper or reasonable manner" [5]. Each of these definitions refers to fixed values, namely non-discrimination, favouritism and equal distribution. Important potential issues to consider are related to the methods or mechanisms for ensuring equity and the choice of parameters to be shared. Fairness can be seen from many different perspectives. Among the others, fairness can be considered from a supplier's or user's perspective, from a hardware or software level, or in terms of QoS or QoE (Fig. 1).

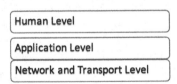

Fig. 1. Levels of fairness [6]

In [6], fairness was divided into three different levels of consideration: the network and transport level, the application level and the human level. The lowest level of fairness, i.e., the network layer, affects the fairness of the whole system and therefore it is so important to ensure fair distribution at this level. The network layer is responsible for major number of important issues such as the transmission priorities, the level of network congestion or the transmission route. At the transport level, the reliability of transmission is ensured. In addition, at this level, correlations are checked and error tracking, multiplexing and flow correction are performed. At the network level, communication protocols such as Transmission Control Protocol (TCP) or User Datagram Protocol (UDP) operate. TCP is a connection protocol that ensures reliable data delivery. It is used for video streaming using the DASH standard. In this case, it is more important for the user to have no interruptions than the short duration of transmission delivery. UDP, on the other hand, is a connectionless protocol. In this case, there is no acknowledgement of packet receipt, and the packets order is not followed. This protocol is most often used for live streaming, due to the high need to ensure no delays in the delivery of the transmission. Currently, both these protocols are used simultaneously, which creates some problems with fair bandwidth distribution. In [7], it was shown that due to the lack of traffic control mechanisms in the UDP protocol, there is significantly more traffic generated from the UDP protocol than TCP when both protocols are used simultaneously. This results in greater latency and even a complete suspension of the traffic generated by TCP. An attempt to ensure fairness by implementing queuing systems was presented in [8].

The application layer is used, among others, to identify communication partners, determine resource availability, synchronize communication, transfer files, process information, or manage files. Fairness at the horizontal level is affected by the interface issued for the purpose of user communication with the embedded network services. It was noted in [9] that providing fairness at the network level does not provide fairness at the application level. It has been shown that an equal bandwidth range at the network level results in different levels of end-user experience (QoE) quality.

The third level at which fairness can be considered is the human level. At this level, fairness between two users or between a user and a computer/robot is considered. The need for fairness between a computer/robot and a second computer/robot should also be taken into account together with the technology stamp. In the Internet of Things which is currently rapidly developing, this issue is even more important because of the amount of data which need to be transmitted. A proposal for a fair distribution between two computers using a connection schedule is made in [10].

As already mentioned, fairness can be considered in terms of quality of service or quality of experience. In [11], QoS is defined as a set of technical and other parameters of a system that control its functionality and need to be adapted based on user satisfaction. In addition, QoS management in the context of distributed multimedia systems, sets the appropriate parameters and reserves the necessary resources to achieve the required functionality and optimize the performance of the entire system. Depending on the services provided by the system, there are several customizable sets of parameters that characterize the operation of the system, i.e. parameters related to performance, format, synchronization, system economy, etc., which can be adjusted to meet the needs of the system. According to [12], QoE is the degree of satisfaction or dissatisfaction of a user related to the application or service. The level of satisfaction (or dissatisfaction) is the consequence of meeting the user's expectations regarding the usefulness of using the application or service, taking into account the user's personality and current state of the user.

The most popular and commonly used measure of fairness is the one proposed by Jain [13]. This measure determines fairness as the ratio of the square of the sum of the resources allocated to each user divided by the sum of the squares of those resources. This measure is used to determine fairness in terms of QoS. Other measures that also serve to determine QoS fairness are described in [14], where the authors proposed a measure that achieves all the objectives of Jaine's measure of fairness and, at the same time, provides better results in situations that are completely unfair. Another measure of QoS fairness is an entropy described in [15]. The main problem with this measure is that it cannot be applied to one of the users if no resource was assigned. Another measure of fairness for QoS is proposed in [16]. This measure very strictly approaches the allocation of a zero resource to any user, treating this situation as extremely unfair.

All mentioned measures are normalized, monotonous, continuous, intuitive – the values taken are from 0 to 1, where 0 means extremely unfair situation, while 1 is perfectly fair, and independent from the number of users [13].

In [17] and [18], Hossfeld defined fairness in terms of the quality of experience and proposed a fairness measure based on the experience of the end user. To use Jain's measure of fairness in relation to quality of service [13], Hossfeld added three requirements for fairness in relation to quality of experience. Determining QoE fairness using the Hossfeld formula requires that certain selected values of service quality must be mapped to the quality of experience.

Martinez-Yelmo in [19] presented how QoE should be measured, described the need for a clear definition of measures and the purpose of QoE determination by presenting and comparing different characteristics affecting QoE. The video QoE metric is presented in [9] as a function of the screen size, resolution and viewing distance. Based on [19], and [9], it can be noticed how many different parameters affect the quality of the service perceived by the users.

1.2   Motivation and Goal

Providers and users rate differently the performance of the application they use or distribute. Providers most often use service quality parameters such as bandwidth, latency, and loss rate. However, users are less interested in technical parameters but more in subjective perception, i.e., QoE [20]. Users expect good perceptual quality, which can be achieved on the basis of many factors, including not only technical parameters, but also user experience [21].

In [22], the attention was drawn to the growing problem of emerging network congestion caused by the increasing traffic in high-definition audiovisual content and the lack of sufficient control over network traffic. This causes image quality to decrease and even freeze the frame rate of playback. An important aspect in managing this type of traffic is to ensure the quality of experience and to ensure fairness in this context for various devices in the network.

One of the most sensitive categories of multimedia concerning the lack of fairness are online games. Users do not want to spend time playing a game if noticing unequal opportunities between players. There are many papers showing the impact of service quality parameters (such as network latency, jitter, loss of packets [23–25]) on computer games. However, there is a lack of mechanisms to ensure fairness in games as a whole, not just single technical parameters.

Taking into account today's needs, i.e., increased focus on the experiences and feelings of users, in this paper we propose an algorithm ensuring fair bandwidth distribution in the context of QoE for satisfied users. At the same time, to ensure the connection to the internet for all users, unsatisfied users are allocated non-zero bandwidth while maintaining its fair distribution in relation to QoS for unsatisfied users. In this paper, we show that for the proposed algorithm there are definitely more satisfied customers, and thus less resignation from the service provider are observed in comparison to the commonly used max-min algorithm.

The structure of this paper is as follows. Section 2 discusses the existing solutions and their weaknesses in relation to today's expectations. Section 3 discusses the use of measures to determine fairness. Section 4 presents the proposed algorithm. In Sect. 5 there is an example of the application of the proposed and the max-min algorithms. Section 6 presents the evaluation results obtained after applying the proposed algorithm and the max-min algorithm. Section 7 contains a short summary and Sect. 8 contains suggestions for further work to be done.

2   Related Work

The most popular algorithm to ensure fairness is the max-min algorithm used to maximize the minimal assignment of resources to users, referred to as *max-min fairness* [26]. This algorithm starts with zero resource allocation for all nodes and then increases resource allocation for all nodes until the link is saturated or the full resource is used. The final effect of the max-min algorithm is a full division of the resource in such a way that users with less resource than others have full capacity for this resource [27]. Another popular algorithm is an algorithm that provides proportional fairness [28].

As can be noticed, the highest value of fairness based on the measure in [13], will occur when all users get the same equal resource allocation or some users don't get any allocation and the rest of the users get an equal allocation. Not receiving any allocation is clearly and intuitively unfair. Equal resource allocation does not results in an equal range of user experience, because each user has different hardware, which leads to different levels of expected user experience [9]. In [29], equal flows were rejected as a solution that is unfair. Since the user's decision to quit is based on feelings and measures of subjective rather than objective technical parameters, an equal distribution of service quality parameters will not be beneficial to service providers.

Current solutions for video transmission use Adaptive BitRate (ABR) algorithms like MPEG-DASH [30,31], which dynamically selects the bit rate and resolution of video to provide the best possible QoE in a given situation, taking into account the limitations of the user's device or the maximum bandwidth of the network. ABR algorithms work well to improve individual QoE of users, which is good when the network is not overloaded. When amount of network resources is lower than the needs reported by users, the use of ABR algorithms leads to favouring some customers over the others. The reason for the unequal distribution of QoE among customers is that they use equal bandwidth distribution mechanisms without taking into account the potentially different user devices [32].

The dynamic allocation of resources is shown in [33], based on Software Defined Networks (SDN), to manage DASH networks and similar solutions such as Apple HLS. This solution supports various algorithms for adapting the customer's bandwidth and limits the customer's choice of bandwidth and data quality. These limitations are used to optimize the distribution of bandwidth between customers and are communicated to customers as recommendations. The customer uses the recommendation as an upper bandwidth limit and sets the buffer level during the process of selecting the target bandwidth. The process of selecting the bandwidth for each client takes into account the available bandwidth, buffer occupation, video content type constraints and device capabilities. However, as can be seen, this is the optimal solution in terms of ensuring the maximum number of users, but it does not provide a fair solution.

From the above considerations, a number of key problems can be drawn in the current market solutions, namely:

- Current market solutions focus more on the equal distribution of technical parameters than on the quality of experience. For this reason, users with the same allocated bandwidth but completely different equipment will have a different experience.
- Solutions based on ABR algorithms do not provide an equal level of QoE,
- SDN-based solutions do not provide a satisfactory level of QoE.

The proposed algorithm focuses on:

- Ensuring fairness with regards to the quality of the experience so that users can fairly share the resource and have a fair chance, for example in a computer gaming environment or accessing to a single subnet.
- Ensuring a certain minimum quality of experience, ensuring a maximum number of users who will remain with the service provider.

To the best of our knowledge, there is no mechanism in the market yet that would ensure fairness while minimising the churn. Therefore, the proposed algorithm can become a very attractive solution for providers.

3   Preliminaries

The proposed algorithm is based on the division of n end users into m satisfied and k unsatisfied. The subjective opinion of users is expressed by the mean subjective score (MOS) [34]. A satisfied user is i-th user whose subjective opinion MOS_i is above the minimum subjective opinion MOS_{min} specified by the provider.

3.1   Fairness of Satisfied Users

Hossfeld in papers [17] and [18] included a comprehensive description of the fairness index F, including the use of web downloads in simple web research. He also used this research to compare the proposed measure with popular Jain fairness measure [13]. Hossfeld Fairness Index was developed with the intention of measuring fairness in the context of the quality of experience QoE. To use Hossfeld's measurements it is necessary to first map the quality of service QoS to the quality of experience QoE [18]. [35] presented different ways of mapping QoS indices to QoE.

Hossfeld fairness index proposed in [17] is defined as:

$$F = 1 - \frac{2\sigma}{H - L} \tag{1}$$

where: $\sigma^2 = \frac{1}{n-1}\sum_{i=1}^{n}(Y_i - \mu)^2$, $\mu = \frac{\sum_{i=1}^{n}Y_i}{n}$, H – the maximum value of the parameter, L – the minimum value of the parameter, Y_i – QoE parameter, i – user index, n – number of users

In this paper the level of fairness for satisfied users F is determined by the proposed formula according to Hossfeld. The subjective opinion expressed by the mean subjective score MOS_i is used as a parameter $QoE(Y)$ on which depends the fairness index so $Y_i = MOS_i$. The highest MOS_i value is 5 (H), the lowest is 1 (L). On the basis of the research conducted by Advancing Customer Experience (ACE), a formula reflecting the subjective opinion of the MOS_i user was determined, depending on the standardized file size fn_i downloaded by the i user and the bandwidth bw_i user i [36]:

$$MOS_i = \frac{0.755}{\sqrt{fn_i}} \cdot ln(bw_i) + 1.268 \tag{2}$$

where: MOS_i – subjective opinion of user i, bw_i – bandwidth allocated to the user i, fn_i – normalised size of the file to be downloaded by the user i.

3.2   Fairness of Unsatisfied Users

This paper uses the S fairness measure proposed in [16]. This measure was used to ensure fairness to QoS for unsatisfied users. The measure takes values from zero to one, with zero being extremely unfair, and one being perfectly fair. This measure is used in the proposed mechanism because it takes a very strict approach to a situation where any end-user would get a zero bandwidth allocation treating this situation as extremely unfair. The proposed mechanism ensures that there will be no end-user zero bandwidth allocation, thus ensuring an internet connection for all users.

$$S = k^k \cdot \prod_{i=1}^{k}\left(\frac{x_i}{\sum_{j=1}^{k}x_j}\right), \tag{3}$$

where: S – measure of fairness for unsatisfied users, k – number of unsatisfied users, x – user resource. In the proposed algorithm the size of the assigned bandwidth is taken as the resource, i.e. $x_i = bw_i$.

3.3 Assumptions

The purpose of the proposed algorithm is to ensure a minimum number of service resignations, i.e., to minimize the phenomenon of churn, while maintaining an internet connection for all users and ensuring QoE fairness among satisfied users and QoS fairness among unsatisfied users. The proposed algorithm allows service providers to regulate the value of bandwidth allocations for unsatisfied users using a service provider ratio. It is important to allocate non-zero bandwidth to unsatisfied users to maintain internet connection for all users.

The following terms are used in this paper:

- MOS - subjective user opinion is expressed by the mean subjective score
- MOS_{min} - minimum subjective opinion is specified by the provider.
- number of satisfied users m—users whose subjective opinion is higher than or equal to MOS_{min}, due to which they are satisfied with the service and will not resign from it,
- number of unsatisfied users k—users whose subjective opinion is lower than MOS_{min}, thus they are unsatisfied with the service and will resign from it,
- the provider coefficient q—the coefficient in the range above zero to one, is used to determine the value of the allocated bandwidth for unsatisfied users bw_i. The higher the value of the provider coefficient, the higher the value of the bandwidth allocated to unsatisfied users according to the formula (7) will be,
- MOS_{mean} – arithmetic mean of subjective user opinions satisfied.

4   Proposed Algorithm

The proposed algorithm of the distribution of resources guarantees the maximum number of satisfied users at a given value of the service provider's coefficient. The current solutions are focused on ensuring fairness in QoS aspect and ignore the opinion and experience of the end user. The proposed algorithm focuses on maximising the number of satisfied users. It ensures fairness in terms of QoE to satisfied users and fairness in terms of QoS to unsatisfied users.

For satisfied users, it is important to be fair in QoE aspect, while for unsatisfied users who give up services, it is important that the distribution of the link bandwidth among them is fair in relation to QoS. In addition, it must be ensured that unsatisfied users are not allocated zero bandwidth ($bw_{min} > 0$), because such a situation is seen as extremely unfair [16] as some users have no connection to the internet. Service providers want the number of unsatisfied customers to be as low as possible and the subjective average opinion as high as possible, as this attracts new customers.

4.1   The Main Parts of the Proposed Algorithm

The proposed algorithm can be divided into two independent Algorithms 1 and 2. Algorithm 2 can be divided into two parts 2a and 2b, according to different maximized values:

Algorithm 1. Maximising the number of users satisfied with the service.

$$M_{countSatisfied} = \frac{m}{n} \tag{4}$$

Algorithm 2a. Maximize the fairness index with the maximum mean subjective opinion for satisfied users. (Maximising QoE fairness).

$$M_{fairnessSatisfied} = \frac{F_m \cdot MOS_{mean}}{H} \tag{5}$$

Algorithm 2b. Maximizing the fairness measure for unsatisfied users. (Maximising QoS fairness).

$$M_{fairnessUnsatisfied} = S_k \tag{6}$$

where: m – number of satisfied users, n – number of all users, F_m – Hossfeld Fairness Index for satisfied users, MOS_{mean} – mean opinion from satisfied users, H – maximum possible subjective opinion, S_k – Nowicki Fairness Index for unsatisfied users

4.2   Scheme of the Algorithm

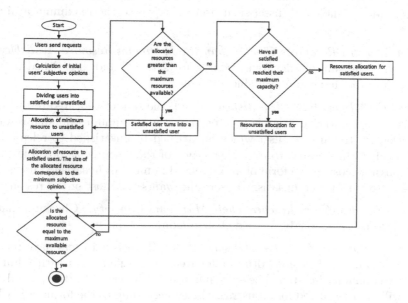

Fig. 2. Algorithm of resource allocation.

Users Send Requests: End users send to the computing unit the required bandwidth and file size they want to download (Fig. 2).

Calculation of Initial Users' Subjective Opinions: The computing unit calculates the preliminary subjective opinion of the users on the basis of the submitted applications (according to the relation (2)).

Division of Users Into Satisfied and Unsatisfied: Based on opinions set out in the previous stage, users are divided into those who are satisfied and those who are unsatisfied. Unsatisfied user is user whose subjective opinion is lower than MOS_{min}. MOS_{min} is determined by the provider.

Allocation of Minimum Resource to Unsatisfied Users: Unsatisfied users shall be allocated a minimum bandwidth, in accordance with work [37], as defined by the formula:

$$bwTemp_l = q \cdot bwTemp_{minUnsatisfied} \tag{7}$$

where: l – unsatisfied user $(l = 1, 2, 3 ..., k), q$ – service provider's coefficient $q \in (0, 1>, bwTemp_{minUnsatisfied} = min(bw_{min}; \frac{BW}{n})$, bw_{min} – minimum request among unsatisfied users.

Allocation of Resource to Satisfied Users. The Size of the Allocated Resource Corresponds to the Minimum Subjective Opinion: The bandwidth will be determined for all satisfied users according to the formula in accordance with [37]. For this stage $MOS = MOS_{min}$

$$bwTemp_i = min(e^{\frac{((MOS-1,268) \cdot \sqrt{fn_i})}{0,775}}; bw_i) \tag{8}$$

Is the Resource Fully Allocated?: It is verified whether the sum of allocated resources for satisfied and dissatisfied users is equal to the maximum available resource.

Are the Allocated Resources Greater than the Maximum Resource Available?: It is verified if the sum of allocated resource for satisfied and dissatisfied users is greater than the maximum available resource.

Satisfied User Turns Into a Unsatisfied User: If the sum of allocated resource for satisfied and dissatisfied users is greater than the maximum available resource it is necessary to turn one user which is satisfied to unsatisfied to release allocated bandwidth. In order to release as much of the resource as possible at the lowest possible cost in the form of an increased number of unsatisfied users, it is necessary to turn into a unsatisfied user who wants the most of the resource.

Have All Satisfied Users Reached Their Maximum Capacity?: Checking that all satisfied users have already reached their maximum capacity.

Resources Allocation for Satisfied Users: If not all satisfied users have reached maximum bandwidth, bandwidth is allocated to satisfied users with minimum subjective opinion plus step. The step increases with every iteration. The bandwidth will be determined for all users satisfied according to the formula 8, where $MOS = MOS_{min} + step, step = 0.1 \cdot$ iteration.

Resources Allocation for Unsatisfied Users: If the bandwidth is distributed as above and the sum of the allocated bandwidth of all users is less than the maximum possible network bandwidth, and all satisfied users have reached their maximum bandwidth, the remaining part of the bandwidth is distributed among unsatisfied users. This will not change the unsatisfied user into the satisfied user as it is impossible, but will maximize the usage of the entire network resources.

5   Max-Min vs Proposed Algorithm

In this section, we present the distribution of bandwidth between 5 users using max-min algorithm and the proposed solution. Table 1 shows the users' requests – max bandwidth (bw_i) that the user can access (i.e., what users want) and the size of the downloaded file f_{size}.

Table 1. The requests for individual nodes

User	Request	
	File size [Mb]	Bandwidth [Mb/s]
1	1	5
2	3	10
3	8	2
4	4	12
5	1	3

Assumptions:

- minimum subjective user opinion: $MOS_{min} = 3.0$,
- the service provider's coefficient: $q = 1.0$,
- maximum bandwidth of the whole network: $BW = 15$.

5.1   Step 1

Table 2. Bandwidth allocation in first step for the proposed and max-min algorithms

User index	BW [Mb/s] – Step 1	
	Proposed algorithm	Max-Min algorithm
1	2.79	2.00
2	3.51	2.00
3	2.00	2.00
4	3.76	2.00
5	2.00	2.00

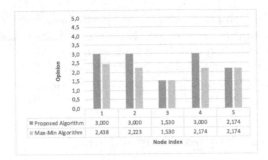

Fig. 3. First stage of bandwidth distribution for the proposed and max-min algorithms.

In max-min algorithm all users are assigned a bandwidth value equal to the lowest user request – Table 2. Then it is checked whether the entire bandwidth has been distributed. In this example, the remaining part of the bandwidth is unallocated. So there is another iteration of bandwidth distribution.

The proposed algorithm uses the information from the submitted requests and on their basis it is determined that user 3 and 5 will always be unsatisfied – Fig. 3. Even if they receive the bandwidth required by them, these users will have a subjective opinion lower than MOS_{min} so they will be unsatisfied and will resign. For this reason, from the point of view of the service provider there is no benefit to give them more bandwidth than is minimally allowed (users will resign anyway). For this reason users 3 and 5 have been assigned bandwidth equal to 2 (i.e., the minimum allowed bandwidth assignment value determined according to the formula (7)) – Table 2. The other users were assigned the bandwidth in such a amount (according to formula (8)) that their subjective opinion was equal to MOS_{min}, i.e., 3.0.

5.2   Step 2

According to the max-min algorithm, bandwidth allocation is increased for each user until one or more nodes reach their maximum value or the entire amount of available resources is distributed. In proposed algorithm, the subjective opinions of satisfied users shall be increased and the bandwidth allocated to satisfied users until the entire bandwidth is distributed (Fig. 4 and Table 3).

5.3   Step 3 – Final Step

The final distribution of the resource after applying the max-min algorithm is shown in Fig. 5. Assuming the opinion MOS_{min} equals to 3.0 (the minimum opinion below which the user will resign) 4 out of 5 users will resign from the services of the provider after applying the max-min algorithm. In this case, the churn resignation rate is 80%. After applying the proposed algorithm, the number of satisfied users is 3 out of 5. Thus, the churn resignation rate is 40%.

Table 3. Bandwidth allocation in second step for the proposed and max-min algorithms

User index	BW [Mb/s] – Step 2	
	Proposed algorithm	Max-Min algorithm
1	2.96	3.00
2	3.78	3.00
3	2.00	2.00
4	3.44	3.00
5	2.00	3.00

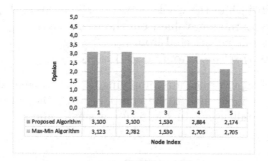

Fig. 4. Second stage of bandwidth distribution for proposed and max-min algorithms.

Table 4. Bandwidth allocation in last step for the proposed and max-min algorithms

User index	BW [Mb/s] – Step 3	
	Proposed algorithm	Max-Min algorithm
1	3.01	3.33
2	3.85	3.33
3	2.00	2.00
4	4.14	3.33
5	2.00	3.00

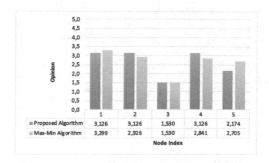

Fig. 5. Final bandwidth distribution between users for both algorithms.

It is definitely more beneficial from the point of view of service providers than after using the max-min algorithm (Table 4).

As can be seen, both the proposed algorithm and the max-min algorithm have fully allocated the available bandwidth.

6   Evaluation Results

Assumptions:

- minimum subjective user opinion: $MOS_{min} = 3.0$,
- the service provider's coefficient: $q = 1.0$,
- maximum bandwidth of the whole network: $BW = 1000\,\text{Mb/s}$
- number of users n: random value from the range $(3, 750)$
- bandwidth requests bw_i selected from $(2, 10)$ range
- file size requests f_i selected from $(1, 100)$ range

Calculations were performed after applying the proposed algorithm and max-min algorithm. The first value to be compared was the ratio of the number of satisfied users to all users. In Fig. 6, for a small number of users for both

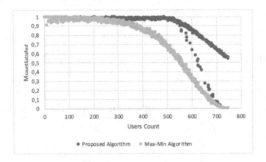

Fig. 6. $M_{countSatisfied}$ value depending on the number of users for the proposed algorithm and max-min algorithm.

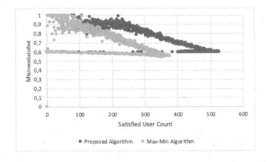

Fig. 7. $M_{fairnessSatisfied}$ value depending on the number of users for the proposed algorithm and max-min algorithm.

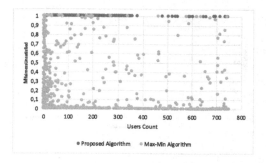

Fig. 8. $M_{fairnessUnsatisfied}$ value is the measure of fairness according to Nowicki for unsatisfied users after using the proposed algorithm and max-min algorithm.

algorithms the number of users is close to the number of all users. This is due to the non-overloaded network scenario, i.e. the users got the bandwidth allocations they requested. Only after the occurrence of network saturation – about 300 users – the different mechanisms of bandwidth allocations begin to work and both curves in Fig. 6 start to deviate from each other (Fig.7).

The next value to be compared is the measure of fairness of unsatisfied users The results are shown in Fig. 8. For all attempts, the values of Nowicki fairness index for unsatisfied users are close to 1. For the max-min algorithm, the values are close to 0.

In Fig. 9, it can be seen that for max-min algorithm the number of satisfied users reaches 380, as it is the maximum number of satisfied users that occurs for the calculated cases for the max-min algorithm. In case when proposed algorithm is used, the curve is able to reach about 520 users. As can be seen, this is a significantly higher number of satisfied users. Additionally, the fairness values for the max-min algorithm never take higher values than for the proposed algorithm.

As can be seen from Fig. 10, the ratio of the number of unsatisfied users to all users is much higher after using the max-min algorithm than the proposed one. It directly affects also the number of resignations, so after applying the proposed algorithm the number of resignations is much lower.

An important factor influencing the size of the fairness indexes of the whole system is the value of the allocated bandwidth for unsatisfied users. In the proposed algorithm, the amount of allocated bandwidth for unsatisfied users may be decreased or increased by the service provider's coefficient q (see formula 7) (Figs. 11 and 12).

With the increase of the value of service provider coefficient, Hossfeld's fairness index for the entire system increases but the number of satisfied users decreases. In the proposed algorithm, the service provider can decide whether to increase the number of users satisfied at the expense of the fairness of the whole system.

Figures 13, 14 and 15 show the nature for different MOS_{min} values. As can be noticed charts are similar, shifted relative to the X axis. The increase of the number of unsatisfied users is similar in all three charts, both for the pro-

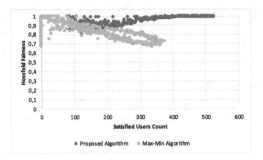

Fig. 9. The dependence of the fairness index for satisfied users on the number of users for the proposed algorithm and max-min algorithm.

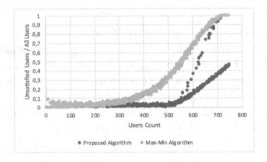

Fig. 10. The relation between the number of unsatisfied users and all users to the number of users for the proposed algorithm and the max-min algorithm.

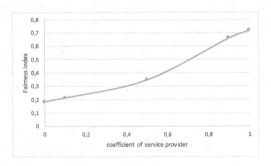

Fig. 11. Hossfeld Fairness Index of all users depending on the service provider coefficient q.

posed algorithm (increase of about 0.2 for 100 users) and the max-min algorithm (increase of about 0.4 for 100 users). With a higher MOS_{min} value, an increase in the ratio of unsatisfied users to all users occurs with fewer users. This is due to the need to allocate more bandwidth to users so that users will be satisfied with a higher MOS_{min}. Thus, the same amount of bandwidth with a higher MOS_{min} will be distributed to fewer users than with a lower MOS_{min}.

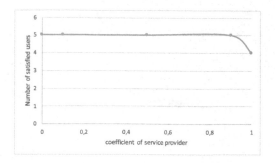

Fig. 12. Number of satisfied users depending on the service provider's coefficient q.

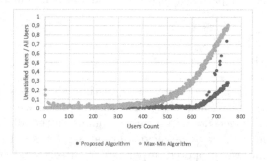

Fig. 13. The relation between the number of unsatisfied users and all users to the number of users for the proposed algorithm and the max-min algorithm for min $MOS_{min} = 2.5$.

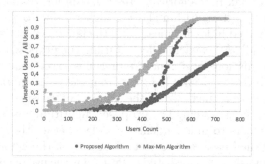

Fig. 14. The relation between the number of unsatisfied users and all users to the number of users for the proposed algorithm and the max-min algorithm for min $MOS_{min} = 3.75$.

A change in the MOS_{min} value does not affect the distribution of the assumed values of the index of Hossfeld fairness. For all calculated values the distribution looks like Fig. 9. The decision on the level of the MOS_{min} value is left to the service provider as they know best what is the minimum user opinion at which the user decides to stay with the service provider.

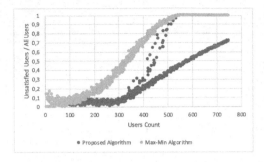

Fig. 15. The relation between the number of unsatisfied users and all users to the number of users for the proposed algorithm and the max-min algorithm for min $MOS_{min} = 4.5$.

7 Conclusions

To meet today's needs, we proposed the algorithm which maximizes the experience of fairness among satisfied users, their mean opinion while also maximizing fairness among unsatisfied users in terms of equalizing the allocated bandwidth, and maximizes the number of satisfied users at a given service provider coefficient. The algorithm was proposed in particular for services where there is high competition and the need to minimize the number of resignations. An examples of such services are systems which allow users to download music or movies files.

In the context of the QoE, the proposed algorithm works much better than the commonly used max-min algorithm. In each computational example given in this paper, the proposed algorithm showed higher values of fairness indices (in particular the dedicated index for measuring the fairness in terms of QoE - index proposed by Hossfeld). A beneficial effect of using the proposed algorithm is that the number of resignations was almost four times lower than for the max-min algorithm, which is a very desirable factor from the point of view of suppliers.

It is also showed that maximizing the number of satisfied users may reduce the fairness of the whole system. The proposed algorithm has a mechanism that allows service providers to independently decide whether they want to increase the number of satisfied users at the cost of fairness of the whole system.

The difficulty in using the presented algorithm is the use of correct mapping of measurable QoS parameters to perceptible QoE parameters. This is complicated due to the fact that for each service this mapping is different, often not yet defined.

8 Further Works

In this paper, we made a research to determine the representation of the subjective opinion of the user while using the services: downloading files and browsing websites in a wireless network environment [36]. To use the presented algorithm for wired networks, it is necessary to do the experiments to determine the constants a and b in the formula reflecting the bandwidth of the user's subjective opinion. The test should be a repetition of experiments from [36] and [38] in the wired network environment.

References

1. White paper: Cisco visual networking index: forecast and trends, 2017–2022 (2018)
2. Spiteri, K., Sitaraman, R., Sparacio, D.: From theory to practice: improving bitrate adaptation in the dash reference player. In: Proceedings of the 9th ACM Multimedia Systems Conference, pp. 123–137 (2018). https://doi.org/10.1145/3204949.3204953
3. Online: Definition of fairness in English by vocabulary dictionaries. https://www.vocabulary.com/dictionary/fairness. Accessed 7 Feb 2020
4. Online: Definition of fairness in English by Oxford dictionaries. https://en.oxforddictionaries.com/definition/fairnes. Accessed 7 Feb 2020
5. Online: Definition of fairness in English by Cambridge dictionaries. https://dictionary.cambridge.org/dictionary/english/fairness. Accessed 7 Feb 2020
6. Basil, A.O., Mu, M., Al-Sherbaz, A.: A software defined network based research on fairness in multimedia. In: Proceedings of the 1st International Workshop on Fairness, Accountability, and Transparency in MultiMedia. FAT/MM 2019, pp. 11–18. ACM (2019) https://doi.org/10.1145/3347447.3356750
7. Gupta, V., Krishnamurthy, S.V., Faloutsos, M.: Improving the performance of TCP in the presence of interacting UDP flows in ad hoc networks. In: Mitrou, N., Kontovasilis, K., Rouskas, G.N., Iliadis, I., Merakos, L. (eds.) NETWORKING 2004. LNCS, vol. 3042, pp. 64–75. Springer, Heidelberg (2004). https://doi.org/10.1007/978-3-540-24693-0_6
8. Domański, A., Domańska, J., Nowak, S., Czachórski, T.: A contribution to the fair scheduling for the TCP and UDP streams. In: Kwiecień, A., Gaj, P., Stera, P. (eds.) CN 2010. CCIS, vol. 79, pp. 207–216. Springer, Heidelberg (2010). https://doi.org/10.1007/978-3-642-13861-4_21
9. Mansy, A., Fayed, M., Ammar, M.: Network-layer fairness for adaptive video streams. In: 2015 IFIP Networking Conference (IFIP Networking), pp. 1–9 (2015). https://doi.org/10.1109/IFIPNetworking.2015.7145310
10. Narayanan, R., Srinivasan, M., Karthikeya, S.A., Murthy, C.S.R.: A novel fairness-driven approach for heterogeneous gateways' link scheduling in IoT networks. In: 2017 IEEE International Conference on Communications (ICC), pp. 1–7 (2017). https://doi.org/10.1109/ICC.2017.7996818
11. Malamos, A.G., Malamas, E.N., Varvarigou, T.A., Ahuja, S.R.: On the definition, modelling, and implementation of quality of service (QoS) in distributed multimedia systems. In: Proceedings of IEEE International Symposium on Computers and Communications (Cat. No. PR00250), pp. 397–403 (1999). https://doi.org/10.1109/ISCC.1999.780929
12. Brunnström, K., Moor, K., Dooms, A., Egger-Lampl, S., et al.: Qualinet White Paper on Definitions of Quality of Experience (2013)
13. Jain, R., Chiu, D.M., WR, H.: A quantitative measure of fairness and discrimination for resource allocation in shared computer systems. CoRR cs.NI/9809099 (1998)

14. Chen, Z., Zhang, C.: A new measurement fornetwork sharing fairness. Comput. Math. Appl. **50**(5), 803–808 (2005)
15. Shannon, C.E.: A mathematical theory of communication. Bell Syst. Tech. J. **27**(3), 379–423 (1948). https://doi.org/10.1002/j.1538-7305.1948.tb01338.x
16. Nowicki, K., Malinowski, A., Sikorski, M.: More just measure of fairness for sharing network resources. In: Gaj, P., Kwiecień, A., Stera, P. (eds.) CN 2016. CCIS, vol. 608, pp. 52–58. Springer, Cham (2016). https://doi.org/10.1007/978-3-319-39207-3_5
17. Hossfeld, T., Skorin-Kapov, L., Heegaard, P.E., Varela, M.: Definition of QoE fairness in shared systems. IEEE Commun. Lett. **21**(1), 184–187 (2017). https://doi.org/10.1109/LCOMM.2016.2616342
18. Hossfeld, T., Skorin-Kapov, L., Heegaard, P.E., Varela, M.: A new QoE fairness index for QoE management. Qual. User Exp. **3**(1), 4 (2018). https://doi.org/10.1007/s41233-018-0017
19. Martinez-Yelmo, I., Seoane, I., Guerrero, C.: Fair Quality of Experience (QoE) measurements related with networking technologies. In: Osipov, E., Kassler, A., Bohnert, T.M., Masip-Bruin, X. (eds.) WWIC 2010. LNCS, vol. 6074, pp. 228–239. Springer, Heidelberg (2010). https://doi.org/10.1007/978-3-642-13315-2_19
20. Shaikh, J., Fiedler, M., Collange, D.: Quality of Experience from user and network perspectives. Ann. Telecommun. Annales des Télécommunications **65**(1), 47–57 (2010). https://doi.org/10.1007/s12243-009-0142
21. Piamrat, K., Viho, C., Bonnin, J., Ksentini, A.: Quality of experience measurements for video streaming over wireless networks. In: 2009 Sixth International Conference on Information Technology: New Generations, pp. 1184–1189 (2009) https://doi.org/10.1109/ITNG.2009.121
22. Georgopoulos, P., Elkhatib, Y., Broadbent, M., Mu, M., Race, N.: Towards network-wide QoE fairness using openflow-assisted adaptive video streaming. In: Proceedings of the 2013 ACM SIGCOMM Workshop on Future Human-centric Multimedia Networking, pp. 15–20 (2013). https://doi.org/10.1145/2491172.2491181
23. Brun, J., Safaei, F., Boustead, P.: Fairness and playability in online multiplayer games. In: 3rd IEEE Consumer Communications and Networking Conference, vol. 2, pp. 1199–1203 (2006). https://doi.org/10.1109/CCNC.2006.1593228
24. Zander, S., Leeder, I., Armitage, G.: Achieving fairness in multiplayer network games through automated latency balancing. In: Proceedings of the 2005 ACM SIGCHI International Conference on Advances in Computer Entertainment Technology, pp. 117–124 (2005). https://doi.org/10.1145/1178477.1178493
25. Hirota, R., icm Kuribayash, S.: Evaluation of fairness in multiplayer network games. In: Proceedings of 2011 IEEE Pacific Rim Conference on Communications, Computers and Signal Processing, pp. 7–11 (2011)
26. Deressa, M., Sheng, M., Wimmers, M., Liu, J., Mekonnen, M.: Maximizing quality of experience in device-to-device communication using an evolutionary algorithm based on users' behavior. IEEE Access **5**, 3878–3888 (2017). https://doi.org/10.1109/ACCESS.2017.2685420
27. Le Boudec, J.Y.: Rate adaptation. A Tutorial, Congestion Control and Fairness (2005)
28. Kushner, H.J., Whiting, P.A.: Convergence of proportional-fair sharing algorithms under general conditions. IEEE Trans. Wirel. Commun. **3**(4), 1250–1259 (2004)
29. Briscoe, B.: Flow rate fairness: dismantling a religion. SIGCOMM Comput. Commun. Rev. **37**(2), 63–74 (2007)

30. Mueller, C., Lederer, S., Timmerer, C.: An evaluation of dynamic adaptive streaming over HTTP in vehicular environments. In: Proceedings of the 4th Workshop on Mobile Video, pp. 37–42. Association for Computing Machinery (2012). https://doi.org/10.1145/2151677.2151686
31. Akhshabi, S., Narayanaswamy, S., Begen, A.C., Dovrolis, C.: An experimental evaluation of rate-adaptive video players over HTTP. Image Commun. **27**(4), 271–287 (2012)
32. Akhtar, Z., et al.: Oboe: auto-tuning video ABR algorithms to network conditions. In: Proceedings of the 2018 Conference of the ACM Special Interest Group on Data Communication, pp. 44–58 (2018). https://doi.org/10.1145/3230543.3230558
33. Bentaleb, A., Begen, A.C., Zimmermann, R.: SDNDASH: improving QoE of HTTP adaptive streaming using software defined networking. In: Proceedings of the 24th ACM International Conference on Multimedia, pp. 1296–1305 (2016). https://doi.org/10.1145/2964284.2964332
34. ITU-T: Recommendation P.800 - Methods for subjective determination of transmission quality. International Telecommunication Union (1996)
35. Hossfeld, T., Heegaard, P.E., Skorin-Kapov, L., Varela, M.: No silver bullet: QoE metrics, QoE fairness, and user diversity in the context of QoE management. In: 2017 Ninth International Conference on Quality of Multimedia Experience (QoMEX), pp. 1–6 (2017). https://doi.org/10.1109/QoMEX.2017.7965671
36. Reichl, P., Egger, S., Schatz, R., D'Alconzo, A.: The logarithmic nature of QoE and the role of the Weber-Fechner law in QoE Assessment. In: 2010 IEEE International Conference on Communications, pp. 1–5 (2010). https://doi.org/10.1109/ICC.2010.5501894
37. Mazur, I.: Ensuring of fairness in high speed computer networks. M.Sc. thesis, University of Technology Gdansk (2018)
38. Egger-Lampl, S., Reichl, P., Hossfeld, T., Schatz, R.: Time is bandwidth? Narrowing the gap between subjective time perception and quality of experience. In: Proceedings of IEEE International Conference on Communications. (2012). https://doi.org/10.1109/ICC.2012.6363769

Queueing Theory and Queuing Networks

On Comparison of Multiserver Systems with Exponential-Pareto Mixture Distribution

Irina Peshkova[1]([✉]) [ID], Evsey Morozov[1,2,3]([✉]) [ID], and Maria Maltseva[1]([✉]) [ID]

[1] Petrozavodsk State University, Lenin str. 33, Petrozavodsk 185910, Russia
`iaminova@petrsu.ru`, `masha.mariam.maltseva@mail.ru`
[2] Institute of Applied Mathematical Research of the Karelian research Centre of RAS, Pushkinskaya str. 11, Petrozavodsk 185910, Russia
`emorozov@karelia.ru`
[3] Moscow Center for Fundamental and Applied Mathematics, Moscow State University, Moscow 119991, Russia

Abstract. Mixture models arise when at least two different distributions of data sets are presented. In this paper, we introduce the upper and lower bounds for the steady-state performance of a multiserver model of the network node, with Exponential-Pareto mixture distribution of service times. We use the failure rate and stochastic comparison techniques together with coupling of random variables to establish some monotonicity properties of the model. These theoretical results are illustrated by numerical simulation of $GI/G/N$ queueing systems.

Keywords: Failure rate comparison · Multiserver system · Queue size estimation · Finite mixture distribution

1 Introduction

Finite mixtures of distributions can be effectively used to model complex stochastic systems through an appropriate choice of its components. Such models are being used extensively for statistical analysis in many real fields. For example, mixture distributions are used in the analysis of lifetime data and in problems related to the modelling of ageing or failure processes, in particular, for the analysis of the failure times of a sample of items of coherent laser used in telecommunication network. Finite mixture densities are also useful in medical and biology research, in artificial neural networks and robustness studies, income analysis [3, 6–8].

This research is supported by the Ministry of Science and Higher Education of the Russian Federation (project no. 05.577.21.0293, unique identifier RFMEFI57718X0293). The research is supported by Russian Foundation for Basic Research, projects No. 19-57-45022, 19-07-00303, 18-07-00156, 18-07-00147.

P. Gaj et al. (Eds.): CN 2020, CCIS 1231, pp. 141–152, 2020.
https://doi.org/10.1007/978-3-030-50719-0_11

Application of the mixture of distributions in the modeling of queueing systems is often motivated because arrival process and/or service times may be actually generated by quite different distributions, in particular, distributions with the so-called *light or heavy tails*. For example, a randomly selected claim in communication network may be from video stream, from online games, or from the audio stream. In internet of things networks some of data are "small and bursty" like a temperature, pressure or light reading from a sensor. Other devices might create huge amounts of data traffic, like a media content. Thus, for data processing time (associated with service time in queueing models) a *mixture model* may work quite well. For insurance portfolio, many small claims and also a few large claims that generate a heavy-tailed distribution is a frequent situation and so the claims distribution can be modeled as a finite mixture of different distributions [8].

To the best of our knowledge, the explicit expressions for the stationary performance measures of mixture models are hardly available [1,2]. It makes bounds and statistical estimates the most relevant tools for analysis and evaluation of the QoS of the system under consideration. In turn, a monotonicity is often a basic property of the stochastic models leading to the corresponding bounds of the target stationary performance measure. In this regard we mention a classic multiserver infinite buffer system $GI/G/N$ in which both the workload process and queue size obey some well-know monotonicity properties with respect to the input intervals and service times, see for instance [1,19]. On the other hand, finite buffer systems also obey a useful monotonicity property provided the so-called *failure rate* function $r(x)$ of the service time distributions satisfy an ordering [4,16,17,19]. Note that the *failure rate ordering* is stronger then *stochastic ordering* and it is especially useful to compare the queueing systems with different inputs, see [19]. In this regard a challenging actual problem is that it is usually not easy to find explicitly conditions allowing to construct stochastic ordering, especially for mixture distributions.

In this note, we develop the monotonicity property of the multiserver system with infinite buffer using the failure rates ordering of the *Exponential-Pareto mixture service time distributions*.

The main contribution of this research is that, using failure rates comparison, we compare the queueing processes in the system with Exponential-Pareto mixture service time distributions with the queueing processes in the systems either with Exponential distribution or with Pareto service time distribution.

In turn, monotonicity properties of various queueing processes is a key element of the method of *regenerative envelops* recently developed by the authors in the papers [9–11]. In this method, the monotonicity properties are used to derive the lower and upper bounds of the performance measures of the original system using minorant and majorant queueing systems which have classic regenerations. This method opens new possibilities in the regenerative estimation of the QoS of complex systems with finite mixture distributions.

The structure of the paper is as follows. In Sect. 2, we define finite mixture distributions and describe some useful *orderings* of the random variables. In

Sect. 3, we develop the failure rate comparison of the Exponential-Pareto distribution with Exponential and Pareto distribution, respectively. This analysis based on the ordering of the failure rate functions is further applied in Sect. 4 to compare the queue sizes in the multiserver systems with infinite buffers. Finally, in Sect. 5, we illustrate obtained theoretical results by numerical simulation.

2   Stochastic Comparison of Random Variables with Finite Mixture Distributions

In this section we discuss finite mixture distributions and different types of ordering of random variables with such distributions.

We say that random variable (r.v.) X is said to follow a *finite mixture distribution* with n components, if the distribution function (d.f.) of X can be written in the form [7]

$$F_X(x) = p_1 F_1(x) + \cdots + p_n F_n(x), \tag{1}$$

where $F_i(x)$ is called the i-th component d.f. of r.v. X_i; the probability p_i is the i-th mixing proportion, such that $\sum_{i=1}^{n} p_i = 1$, $i = 1, \ldots, n$.

For a r.v. X, denote f_X the density, $\mathsf{E}X$ the expectation, and $\overline{F}_X(x) = 1 - F_X(x)$ the tail distribution.

For each x such that the tail distribution $\overline{F}_X(x) = \mathsf{P}(X > x)$ is positive, we can define the *failure rate* function as

$$r_X(x) = \frac{f_X(x)}{\overline{F}_X(x)}. \tag{2}$$

In the *reliability theory or queueing theory*, the quantity $r(x)dx$ can be defined as the condition probability that a failure occurs in the interval of time $(x, x + dx)$ provided that a device is still working at instant x [8].

As the simplest example, we give a two-component d.f. of a r.v. X, and its failure rate function and expectation, respectively,

$$F(x) = pF_1(x) + (1 - p)F_2(x); \tag{3}$$

$$r(x) = \frac{pf_1(x) + (1 - p)f_2(x)}{p\overline{F}_1(x) + (1 - p)\overline{F}_2(x)}; \tag{4}$$

$$\mathsf{E}X = p\mathsf{E}X_1 + (1 - p)\mathsf{E}X_2, \tag{5}$$

where parameter $p \in (0, 1)$.

Now we consider the connection between *stochastic ordering, comparison in failure rate and coupling* of random variables.

Let X and Y be two non-negative r.v.'s with d.f.'s F_X, F_Y, and densities f_X, f_Y, respectively. We say that a r.v. X *is less than* a r.v. Y: *stochastically (in distribution)*, and denote it as $X \underset{st}{\leq} Y$, if

$$\overline{F}_X(x) \leq \overline{F}_Y(x), \ x \geq 0; \tag{6}$$

with probability 1 (w.p.1), denoted $X \leq Y$ *w.p.* 1, if

$$P(X \leq Y) = 1; \tag{7}$$

in *failure rate*, denoted $X \underset{r}{\leq} Y$, if

$$r_X(x) \geq r_Y(x), \ x \geq 0. \tag{8}$$

To construct lower/upper bounds for the steady-state performance of the queue-ing systems, we apply well known relation between stochastic ordering and failure rate ordering [12]:

$$X \underset{r}{\leq} Y \text{ implies } X \underset{st}{\leq} Y \text{ implies } \mathsf{E}X \leq \mathsf{E}Y. \tag{9}$$

In the performance and stability analysis the stochastic ordering plays often a key role. The main reason is that, by coupling technique, stochastic ordering is nicely related to ordering in probability. We recall this statement (for more detail see [18]). Let $\tilde{X}$ be a (stochastic) copy of a r.v. X, i.e., $X \underset{st}{=} \tilde{X}$. A *coupling* of collection of the r.v.'s X_i $i \in M$ is a family of the r.v.'s $\{\tilde{X}_i, i \in M\}$ such that

$$X_i \underset{st}{=} \tilde{X}_i \text{ for all } i \in M,$$

and the r.v.'s $\{X_i\}$ are defined on the *same probability space*, where M is an index set. Now denote F^{-1} the *quantile function* defined as

$$F^{-1}(u) = \inf\{x \in \mathbf{R} : F(x) \geq u\}, \ u \in [0, 1].$$

Then the r. v. $\tilde{X} = F^{-1}(U)$, where r.v. U is uniformly distributed on $[0,1]$, is a *stochastic copy* of X. There exists the following relation between stochastic ordering and ordering in probability [12,18]:

$$X \underset{st}{\leq} Y \text{ if and only if } F_X^{-1}(u) \leq F_Y^{-1}(u), \ u \in [0, 1]. \tag{10}$$

In turn, relation (10) holds if and only if there exists such a coupling $(\tilde{X}; \tilde{Y})$ that

$$\tilde{X} \leq \tilde{Y} \text{ w.p. } 1. \tag{11}$$

Now we consider the comparison of two r.v.'s with different finite mixture dis-tributions. Suppose that

$$X_1 \underset{st}{\leq} X_2 \ and \ Y_1 \underset{st}{\leq} Y_2. \tag{12}$$

Based on the mixture representation (3), we let

$$F_{Z_1} = pF_{X_1} + (1-p)F_{Y_1};$$
$$F_{Z_2} = pF_{X_2} + (1-p)F_{Y_2},$$

where Z_i are r.v.'s corresponding to the distributions F_{Z_i}, $i = 1, 2$.

It is known that if (X_1, X_2), (Y_1, Y_2) are mutually independent, then implies [8]

$$Z_1 \underset{st}{\leq} Z_2, \tag{13}$$

see (12).

The results presented in (9)–(13), in particular, allow to establish monotonicity properties of the stochastic processes in the system by performing sample paths comparison. We present such a result in Sect. 4 for the Exponential-Pareto mixture distribution. A key observation is the following: while a direct *stochastic comparison* of a required performance measure of two different models is often quite complicated, at the same time, the *failure rate ordering*, which indeed implies stochastic ordering, can provide a more straightforward comparison of the target measures. Thus, in what follows, we use the failure rate ordering to establish stochastic ordering of queueing systems as a consequence.

3 Failure Rate Comparison of Exponential-Pareto Mixture Distribution

In this Section, we demonstrate the failure rate ordering technique for mixture distributions. To do this, we consider Exponential-Pareto mixture distribution and establish some auxiliary results used further in Sect. 4.

Let r.v. X have exponential distribution, denoted $Exp(\lambda)$, with d.f.

$$F_X(x) = 1 - e^{-\lambda x}, \; x \geq 0, \; \lambda > 0, \tag{14}$$

and constant failure rate $r_X(x) = \lambda$. Further, let r.v. Y have two-parameter Pareto distribution [5], denoted $Pareto(\alpha, x_0)$, with d.f.

$$F_Y(x) = 1 - \left(\frac{x_0}{x_0 + x} \right)^\alpha, \; x \geq 0, \; x_0 > 0, \; \alpha > 0. \tag{15}$$

It is easy to see that the failure rate of Pareto distribution,

$$r_Y(x) = \frac{\alpha}{x_0 + x}, \; x \geq 0, \tag{16}$$

is monotone decreasing and vanishes as $x \to \infty$.

Suppose that the r.v.'s X and Y, satisfying (14) and (15), respectively, are independent. Denote I a random variable independent of the X and Y, and taking the values 0 and 1 only, with the probabilities

$$P(I = 1) = p, \; P(I = 0) = 1 - p.$$

Then it is called that the r.v.

$$Z = I \cdot X + (1 - I) \cdot Y \tag{17}$$

has an *Exponential-Pareto mixture distribution* with tail distribution

$$\bar{F}_Z(x) = p\bar{F}_X(x) + (1-p)\bar{F}_Y(x) = pe^{-\lambda x} + (1-p)\left(\frac{x_0}{x_0+x}\right)^\alpha. \tag{18}$$

We call p *mixing parameter*. Equation (18) shows that the r.v. Z coincides with r.v. X (exponential) with the probability p, and with r.v. Y (Pareto) with the probability $1-p$. It is easy to calculate that the failure rate of the r.v. (17) is given by

$$r_Z(x) = \frac{p\,r_X(x)a(x) + (1-p)\,r_Y(x)}{p\,a(x) + (1-p)}, \tag{19}$$

where

$$a(x) = e^{-\lambda x}\left(1 + \frac{x}{x_0}\right)^\alpha.$$

We note that

$$r_Z(x) \longrightarrow 0 \quad \text{as} \quad x \to \infty.$$

More exactly, since the derivation

$$\frac{dr_Z(x)}{dx} = -(1-p)\frac{pa(x)(r_X(x) - r_Y(x))^2 + r_Y^2(x)/\alpha(pa(x) + (1-p))}{(pa(x) + (1-p))^2} < 0,$$

that is negative for all x, then $r_Z(x)$ is *monotonically decreasing function*. It is easy to check that the following order between the failure rate functions

$$r_Y(x) \le r_Z(x) \le r_X(x), \tag{20}$$

holds if and only if the following relations are satisfied:

$$r_Y(x) \le \sup_{x \ge 0} r_Y(x) = r_Y(0) = \frac{\alpha}{x_0} \le r_X(x) = \lambda. \tag{21}$$

These relations can be reformulated as the following inequality between given parameters of component distributions:

$$\frac{\alpha}{x_0} \le \lambda. \tag{22}$$

In other words, under assumption (22), the r.v.'s X, Z, Y are ordered in failure rate as in (20), and as a consequence, they are ordered stochastically (again see [12]). It is convenient to formulate it as a separate statement:

$$X \underset{r}{\le} Z \underset{r}{\le} Y \text{ implies } X \underset{st}{\le} Z \underset{st}{\le} Y. \tag{23}$$

It is useful to give explicit expression for the mean of Z:

$$\mathsf{E}Z = p\mathsf{E}X + (1-p)\mathsf{E}Y = \frac{p}{\lambda} + \frac{(1-p)x_0}{\alpha - 1}.$$

It is easy to show that, under condition (22), the expectations are ordered

$$EX \leq EZ \leq EY.$$

Figure 1 demonstrates the ordering of the failure rate functions for Pareto, Exponential-Pareto Mixture and exponential distributions, respectively. Whence it follows that these distributions are stochastically ordered under assumption (22).

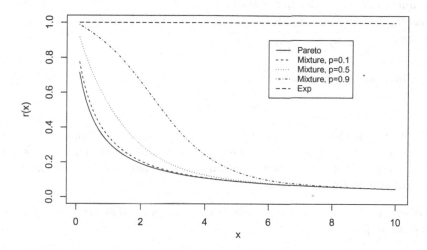

Fig. 1. Failure rate functions of $Exp(1)$, $Pareto(0.5; 0.6)$ and Exponential-Pareto Mixture distributions for a few mixing parameters p.

4 Comparing Multiserver Systems with Different Service Distributions

In this section, we demonstrate how the failure rate comparison allows to stochastically compare the steady-state performance of the multiserver queueing systems with infinite buffers and a renewal input. We consider two queueing systems, denoted by $\Sigma^{(1)}$ and $\Sigma^{(2)}$, with N servers working in parallel. (In what follows, the superscript (i) denotes the number of system.) The service discipline is assumed to be First-Come-First-Served. We denote $S_n^{(i)}$ the service time of customer n, and $t_n^{(i)}$ the arrival instant of customer n. Then the independent identically distributed (iid) interarrival times are defined as

$$\tau_n^{(i)} = t_{n+1}^{(i)} - t_n^{(i)}, \; n \geq 1, \; i = 1, 2.$$

The service times $\{S_n^{(i)}, \; n \geq 1\}$ are assumed to be iid as well, and both sequences are assumed to be independent. Denote $S^{(i)}$ the generic service time, and $\tau^{(i)}$ the generic interarrival time, $i = 1, 2$.

Now we compare the steady-state queue-size processes in the systems $\Sigma^{(1)}$ and $\Sigma^{(2)}$. Denote $Q_n^{(i)}$ the number of customers and

$$W_n^{(i)} = (W_n^{(i1)}, \ldots, W_n^{(iN)}), \quad n \geq 1, \; i = 1, 2,$$

vector of workloads in system i at the arrival instant $t_n^{(i)}$ of the n-th customer. In other words, $W_n^{(ik)}$ is the remaining work to be done by server k of system i at the instant $t_n^{(i)}$, provided the input process is stopped after $t_n^{(i)}$. Denote $\nu_n^{(i)}$ the number of customers which customer n meets in the ith system. By analogy, denote by $Q_n^{(i)}$ the *queue size* in system i. Denote, when exists, the limits (in distribution)

$$Q_n^{(i)} \Rightarrow Q^{(i)}, \; \nu_n^{(i)} \Rightarrow \nu^{(i)}, \; n \to \infty, \quad i = 1, 2.$$

These limits exists, in particular, when

$$\mathsf{E}S^{(i)} < N\mathsf{E}\tau^{(i)}, \tag{24}$$

(recall that N is the number of servers) and the interarrival times $\tau^{(i)}$, $i = 1, 2$ are *non-lattice* [1]. The following statement is a modification of Theorems 4 and 5 in [19], which establish stochastic ordering of the queue sizes and the workloads in the infinite buffer systems.

Theorem 1. *1. Assume the following relations hold in the described $GI/G/N/\infty$ queueing systems*

$$\nu_1^{(1)} \underset{st}{=} \nu_1^{(2)} = 0, \; \tau^{(1)} \underset{st}{=} \tau^{(2)}, \; S^{(1)} \underset{r}{\leq} S^{(2)}. \tag{25}$$

Then, w.p.1,

$$\nu_n^{(1)} \leq \nu_n^{(2)}, \; Q_n^{(1)} \leq Q_n^{(2)}, \; n \geq 1. \tag{26}$$

2. Assume that the following relations hold:

$$W_1^{(1)} \underset{st}{=} W_1^{(2)} = 0, \; \tau^{(1)} \underset{r}{\geq} \tau^{(2)}, \; S^{(1)} \underset{r}{\leq} S^{(2)} \tag{27}$$

Then, w.p.1,

$$W_n^{(1)} \leq W_n^{(2)}, \; n \geq 1. \tag{28}$$

Conditions (25) and (27) seem to be rather restrictive, but indeed they are highly motivated because expressions for comparing failure rates for the most of distributions are analytically available and often have a simple form. Another perhaps a key advantage of these conditions is that they allow to compare queue sizes and workloads in the systems with *different distributions of interarrival and service times*.

5 Numerical Examples

In this section, by numerical simulation, we illustrate for the multiserver systems the comparison results obtained in Sect. 4.

We use *regenerative method* to simulate mutliserver systems, and this approach is one of the most powerful and efficient methods in estimation of the steady-state performance and in the output analysis of complex queueing systems. Briefly, to apply this method, the process describing the systems behavior must possess the following *classical regenerations*: when an arrival meets the system *empty* the future "states" of the system are independent of the pre-history preceding the "empty state". Of course, the queueing systems analyzed in this research have this property. By construction, the trajectory between regeneration points can be divided into iid random elements called *groups*. In turn, this allows further to apply well-developed methods of classical statistics, based on a special form of the Central Limit Theorem (CLT), to construct confidence intervals for the target QoS parameters. At that the CLT is applied to the mentioned iid groups, not to initial data, which in general turn out to be dependent variables. In our experiments presented below, the mean stationary workload and the mean stationary queue size are such QoS parameters. More on regenerative method can be found in [1,2,13–15].

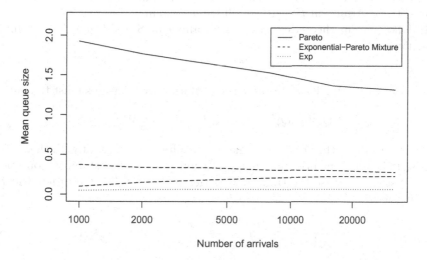

Fig. 2. Lower ($Exp(5)$) and upper ($Pareto(2.5; 0.5)$) bounds of the confidence interval for mean queue size in 4-server system with Exponential-Pareto Mixture distribution of service times.

For the given system $\Sigma^{(2)}$ with Exponential-Pareto mixture distribution of service times (with parameters λ_s, α_s, x_0), we construct two new systems: system $\Sigma^{(1)}$ with an exponential service times $S^{(1)} =_{st} Exp(\lambda_s)$ and $\Sigma^{(3)}$ with

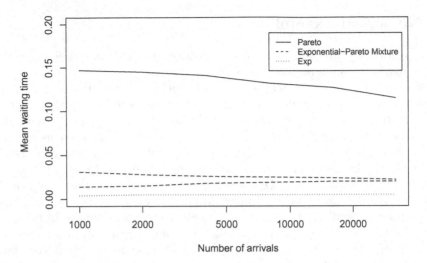

Fig. 3. Lower ($Exp(5)$) and upper ($Pareto(2.5; 0.5)$) bounds of the confidence interval for mean waiting time in 4-server system with Exponential-Pareto Mixture distribution of service times.

Pareto service times $S^{(3)} =_{st} Pareto(\alpha_s, x_0)$. The input flow τ is the same (has exponential distribution $Exp(10)$ in all three systems.

It follows from theoretical results, obtained in Sects. 3 and 4, that under condition

$$\lambda_s \geq \frac{\alpha_s}{x_0}, \tag{29}$$

the following ordering between the basic performance measures must take place:

$$Q_n^{(1)} \leq Q_n^{(2)} \leq Q_n^{(3)} \text{ and } W_n^{(1)} \leq W_n^{(2)} \leq W_n^{(3)}, \ n \geq 1. \tag{30}$$

Indeed, as we show, the obtained numerical results are consistent with theoretical results. It is worth mentioning that condition (29) implies stability conditions of the systems because the traffic intensities $\rho_i = \mathsf{E}S^{(i)}/\mathsf{E}\tau$ are ordered in the following way

$$\rho_1 = \frac{1}{\lambda_s \mathsf{E}\tau} \leq \rho_2 \leq \rho_3 = \frac{x_0}{\mathsf{E}\tau(\alpha_s - 1)} < N, \ \alpha_s > 1, \tag{31}$$

where

$$\rho_2 = \lambda_\tau \left(\frac{p}{\lambda_s} + \frac{(1-p)x_0}{\alpha_s - 1} \right).$$

On the other hand, the requirement $\rho_i < N$ is a stability criterion of N-server system, see (24).

Figure 2 demonstrates lower and upper bounds of the confidence interval for mean queue size for system with 4 servers and Exponential-Pareto Mixture distribution of service times. We compare results of regenerative simulation of this system with the corresponding results for two other systems: the 1st (minorant) system has exponential service time ($Exp(5)$), and the 2nd (majorant) system has Pareto service time ($Pareto(2.5; 0.5)$). Figure 3 shows lower and upper bounds of the confidence interval for mean waiting time in 4-server system with Exponential-Pareto Mixture distribution of service times, with following parameters

$$N = 4, \ \alpha_s = 2.5, \ \lambda_s = 5, \ x_0 = 0.5, \ p = 0.5.$$

The generic interarrival time τ has exponential distribution with rate $\lambda_\tau = 10$. As a result, we obtain

$$\rho_1 = 2 < \rho_2 \approx 2.7 < \rho_3 \approx 3.3.$$

6 Conclusion

In this paper, we study the applicability the *failure rate comparison* of the steady-state performance measures in the multiserver system with Exponential-Pareto mixture service times. We demonstrate a few refined monotonicity results for the systems in which the properties of service time distributions imply specific monotonicity properties of the failure rate functions. Some numerical results based on simulation are also presented. The obtained results may be quite useful for the estimation of the performance measures using *regenerative simulation* of a wide class of queueing systems with mixture distributions. In a future research, the authors in particular, assume to consider sensitivity of upper and lower bounds of performance measures (workload and queueing size) with respect to the mixing parameter and the parameters of Exponential-Pareto mixture distribution of service times. In a future work, the authors aim to study by simulation the systems with a general renewal input.

References

1. Asmussen, S.: Applied Probability and Queues, 2nd edn. Springer, New York (2003). https://doi.org/10.1007/b97236. 439 p.
2. Asmussen, S., Glynn, P.: Stochastic Simulation: Algorithms and Analysis. Springer, New York (2007). https://doi.org/10.1007/978-0-387-69033-9. 476 p.
3. Aven, T., Jensen, U.: Stochastic Models in Reliability. Springer, New York (2013). https://doi.org/10.1007/978-1-4614-7894-2. 297 p.
4. Berger, A., Whitt, W.: Comparisons of multi-server queues with finite waiting rooms. Stoch. Models **8**(4), 719–732 (1992). https://doi.org/10.1080/15326349208807248
5. Goldie, C., Klüppelberg, C.: Subexponential distributions. In: A Practical Guide to Heavy Tails: Statistical Techniques for Analysing Heavy Tails, pp. 435–459. Birkhäuser, Cambridge (1998)
6. Al-Hussaini, E.K., Sultan, K.S.: Reliability and hazard based on finite mixture models. In: Advances in Reliability, Handbook of Statistics, vol. 20, pp. 139–183 (2001). https://doi.org/10.1016/S0169-7161(01)20007-8

7. Mclachlan, G.I., Peel, D.: Finite Mixture Models. Wiley Series in Probability and Statistics. Applied Probability and Statistics Section. Wiley, New York (2001). https://doi.org/10.1002/0471721182

8. Marshall, A., Olkin, I.: Life Distributions: Structure of Nonparametric, Semiparametric and Parametric Families. Springer, New York (2007). https://doi.org/10.1007/978-0-387-68477-2. 783 p.

9. Morozov, E., Peshkova, I., Rumyantsev, A.: On failure rate comparison of finite multiserver systems. In: Vishnevskiy, V.M., Samouylov, K.E., Kozyrev, D.V. (eds.) DCCN 2019. LNCS, vol. 11965, pp. 419–431. Springer, Cham (2019). https://doi.org/10.1007/978-3-030-36614-8_32

10. Morozov, E., Nekrasova, R., Peshkova, I., Rumyantsev, A.: A regeneration-based estimation of high performance multiserver systems. In: Gaj, P., Kwiecień, A., Stera, P. (eds.) CN 2016. CCIS, vol. 608, pp. 271–282. Springer, Cham (2016). https://doi.org/10.1007/978-3-319-39207-3_24

11. Morozov, E., Peshkova, I., Rumyantsev, A.: On regenerative envelopes for cluster model simulation. In: Vishnevskiy, V.M., Samouylov, K.E., Kozyrev, D.V. (eds.) DCCN 2016. CCIS, vol. 678, pp. 222–230. Springer, Cham (2016). https://doi.org/10.1007/978-3-319-51917-3_20

12. Ross, S., Shanthikumar, J.G., Zhu, Z.: On increasing-failure-rate random variables. J. Appl. Probab. 42(3), 797–809 (2005). https://doi.org/10.1239/jap/1127322028

13. Shedler, G.S.: Regeneration and Networks of Queues. Sprigner, New York (1987). https://doi.org/10.1007/978-1-4612-1050-4. 224 p.

14. Shedler, G.S.: Regenerative Stochastic Simulation. Academic Press Inc., Burlington (1992). https://www.elsevier.com/books/regenerative-stochastic-simulation/shedler/978-0-08-092572-1. 400 p.

15. Sigman, K., Wolff, R.W.: A review of regenerative processes. SIAM Rev. 35(2), 269–288 (1993). https://doi.org/10.1137/1035046

16. Sonderman, D.: Comparing multi-server queues with finite waiting rooms, I: same number of servers. Adv. Appl. Probab. 11(2), 439–447 (1979). https://doi.org/10.2307/1426848

17. Sonderman, D.: Comparing multi-server queues with finite waiting rooms, II: different number of servers. Adv. Appl. Probab. 11(2), 448–455 (1979). https://doi.org/10.2307/1426849

18. Thorisson, H.: Coupling, Stationarity, and Regeneration. Springer, New York (2000)

19. Whitt, W.: Comparing counting processes and queues. Adv. Appl. Probab. 13(1), 207–220 (1981). https://doi.org/10.2307/1426475

Unreliable Single-Server Queueing System with Customers of Random Capacity

Oleg Tikhonenko[1] and Marcin Ziółkowski[2(✉)]

[1] Institute of Computer Science, Cardinal Stefan Wyszyński University in Warsaw,
ul. Wóycickiego 1/3, 01-938 Warsaw, Poland
o.tikhonenko@uksw.edu.pl
[2] Institute of Information Technology, Warsaw University of Life Sciences – SGGW,
ul. Nowoursynowska 159, 02-787 Warsaw, Poland
marcin_ziolkowski@sggw.pl

Abstract. In the paper, we investigate one-server queueing system with stationary Poisson arrival process, non-homogeneous customers and unreliable server. As non-homogenity, we mean that each customer is characterized by some arbitrarily distributed random capacity that is called customer volume. Service time of a customer generally depends on his volume. The server can be broken when it is free or busy and the renewal period goes on for random time having an arbitrary distribution. During this period, customers present in the system and arriving to it are not served. Their service continues immediately after renewal period termination. For such systems, we determine the distribution of total volume of customers present in it. An analysis of some special cases and some numerical examples are attached as well.

Keywords: Queueing system with non-homogeneous customers ·
Unreliable queueing system · Total volume · Additional event method ·
Laplace-Stieltjes transform

1 Introduction

A single-server queueing system $M/G/1/\infty$ with unlimited queue is one of the basic models of classical queueing theory and its applications. Many real computer or telecommunication systems satisfy the assumptions of this model: they are composed of only one server, customers arriving to it form Poisson arrival process and service time of a customer is arbitrarily distributed. In some other cases, these assumptions are not strictly satisfied, but behavior of proper systems is very similar (e.g. the queue is limited but long enough or arrival process is close to Poisson one), and we can approximate their characteristics using this model. The results for the classical $M/G/1/\infty$ queueing model are widely known, especially Pollaczek-Khinchine formula for the generating function $P(z) = \sum_{k=0}^{\infty} \mathsf{P}\{\eta = k\}z^k$ of the stationary number of customers η present in the system [2].

© Springer Nature Switzerland AG 2020
P. Gaj et al. (Eds.): CN 2020, CCIS 1231, pp. 153–170, 2020.
https://doi.org/10.1007/978-3-030-50719-0_12

On the other hand, the headway in computer science leads to some new modifications of this model. Indeed, if we focus on analysis of real computer systems, we should take into account the following problems: 1) customers coming to queueing systems are not homogeneous: they usually transport some information (measured in bytes), i.e. different customers have as a rule different volumes (sizes); 2) service time of a customer depends on his volume; 3) servers are unreliable, they can be broken and then must be fixed for some random time. These additional assumptions lead to new queueing models called queueing systems with non-homogeneous customers (assumptions 1, 2) and unreliable servers (assumption 3). The main result for $M/G/1/\infty$ queueing system with non-homogeneous customers, unlimited total volume and customer's service time dependent on his volume is the expression for Laplace-Stieltjes transform $\delta(s) = \int_0^\infty e^{-sx} \mathrm{d}D(x)$ of steady-state customers' total volume σ, where $D(x)$ is distribution function of random variable σ [1,6], and its modifications [9–11,14] that can be treated as generalizations of Pollaczek-Khinchine formula obtained by tools of classical queueing theory [2,12,13].

The main purpose of this paper is an investigation of some modification of $M/G/1/\infty$ model in which we assume that: 1) customers that arrive to the system are characterized by random volume; 2) the server is unreliable, i.e. it can be paused both when it is free or when it is busy; 3) service time of a customer depends on his volume. The analyzed model is the generalization of classical single-server queueing system with unreliable server [3]. For this model, we obtain characteristics of the total volume of customers present in the system for its various versions and investigate some special cases.

The rest of the paper is organized as follows. In the next Sect. 2, we introduce some needed notation and present well known results from the theory of queueing systems with non-homogeneous customers that will help us to analyze the mentioned above model with unreliable server. Then, in Sect. 3, we present the method of additional event [3–5] that is rarely used in papers written in English and its modification for systems under consideration. We also show (as an example) how to use this modification to obtain well-known results for $M/G/1/\infty$ model with non-homogeneous customers and unlimited total volume (these results were obtained earlier with the use of other methods). Section 4 contains some preliminary results for the model with unreliable server. In Sect. 5, we present main statements for the system with unreliable in free state server. To obtain these results, we also use modified method of additional event. We also present here formulae for total volume characteristics (e.g. Laplace-Stieltjes transform of the steady-state total volume and its moments). Section 6 contains analysis of the analogous system, but this time we assume that the server can be also broken if it is busy. In Sect. 7, we present results for some special cases of the model together with numerical examples. The last Sect. 8 contains conclusions and final remarks.

2   Mathematical Description of Systems with Non-homogeneous Customers

We assume that each customer arriving to the system is characterized by some random volume ζ that is a non-negative random variable (RV). Let $\eta(t)$ be the number of customers present in the system at time instant t, $\sigma(t)$ be the sum of the volumes of all these customers (total volume). Our purpose is the determination of random process $\sigma(t)$ characteristics. We also assume that customer's service time ξ generally depends on his capacity ζ. This dependence is determined by the following joint distribution function (DF):

$$F(x,t) = \mathsf{P}\{\zeta < x, \xi < t\}.$$

Let $L(x) = F(x, \infty)$ be DF of customer's volume, $B(t) = F(\infty, t)$ be DF of service time. Let

$$\alpha(s,q) = \mathsf{E}e^{-s\zeta - q\xi} = \int_{x=0}^{\infty} \int_{u=0}^{\infty} e^{-sx - qu} dF(x, u),$$

where $\operatorname{Re} s \geq 0$, $\operatorname{Re} q \geq 0$, be double Laplace-Stieltjes transform (LST) of DF $F(x,t)$. Denote by $\varphi(s) = \alpha(s,0)$ LST of DF $L(x)$ and by $\beta(q) = \alpha(0,q)$ – LST of DF $B(t)$. Let $i, j = 1, 2, \ldots$. Denote by $\Delta(i,j)$ and $\Delta_q(i)$ the following differential operators:

$$\Delta(i,j) = (-1)^{i+j} \frac{\partial^{i+j}}{\partial s^i \partial q^j}, \quad \Delta_q(i) = (-1)^i \frac{\partial^i}{\partial q^i}.$$

Let $\alpha_{i,j} = \Delta(i,j)\alpha(s,q)|_{s=0, q=0}$, $\varphi_i = \Delta_s(i)\varphi(s)|_{s=0}$ and $\beta_i = \Delta_q(i)\beta(q)|_{q=0}$. Then, we have evidently that $\alpha_{i,j}$ is the mixed $(i+j)$th moment of DF $F(x,t)$ and φ_i, β_i are the ith moments of DF $L(x)$ and $B(t)$, respectively (if they exist).

We assume that the arrival process is a stationary Poisson one with parameter a. Assume also that service discipline does not depend on customer's volume ζ and system is empty at the initial time moment $t = 0$, i.e. $\sigma(0) = 0$. Introduce the notation $D(x,t) = \mathsf{P}\{\sigma(t) < x\}$.

Let

$$\overline{\delta}(s,t) = \mathsf{E}e^{-s\sigma(t)} = \int_{x=0}^{\infty} e^{-sx} d_x D(x,t)$$

be LST of the function $D(x,t)$ with respect to x. It is clear that, for arbitrary $t > 0$, the ith moment of the random process $\sigma(t)$ (if it exists) takes the form:

$$\overline{\delta}_i(t) = \mathsf{E}\sigma^i(t) = \Delta_s(i)\overline{\delta}(s,t)\Big|_{s=0}.$$

Denote by

$$\delta(s,q) = \int_0^{\infty} e^{-qt}\overline{\delta}(s,t)dt = \int_0^{\infty} e^{-qt}\mathsf{E}e^{-s\sigma(t)}dt$$

the Laplace transform with respect to t of the function $\overline{\delta}(s,t)$.

Then, we obtain for Laplace transfom $\delta_i(q)$ of $\overline{\delta}_i(t)$ with respect to t that

$$\delta_i(q) = \int_0^\infty e^{-qt}\overline{\delta}_i(t)\mathrm{d}t = \Delta_s(i)\delta(s,q)\Big|_{s=0}.$$

If steady state exists for the system under consideration, i.e. $\sigma(t) \Rightarrow \sigma$ in the sense of a weak convergence, where σ is a steady-state total volume, we can introduce the following steady-state characteristics:

$$D(x) = \mathsf{P}\{\sigma < x\} = \lim_{t\to\infty} D(x,t),$$

$$\delta(s) = \int_0^\infty e^{-sx}\mathrm{d}D(x) = \lim_{t\to\infty} \overline{\delta}(s,t) = \lim_{q\to 0} q\delta(s,q).$$

For steady-state ith moments δ_i of the total volume σ, we obtain:

$$\delta_i = \mathsf{E}\sigma^i = \lim_{t\to\infty} \delta_i(t) = \Delta_s(i)\delta(s)\Big|_{s=0}.$$

Let $\chi(t)$ be the volume of a customer that is served at time instant t. Let $\xi^*(t)$ be the time from the service beginning to the moment t. The next statement was proved in [13].

Lemma 1. *Let* $E_y(x) = \mathsf{P}\{\chi(t) < x \,|\, \xi^*(t) = y\}$ *be conditional DF of the random variable* $\chi(t)$ *under condition* $\xi^*(t) = y$. *Then*

$$dE_y(x) = [1 - B(y)]^{-1} \int_{u=y}^\infty dF(x,u).$$

Hence, the function $E_y(x)$ takes the form:

$$E_y(x) = \mathsf{P}\{\zeta < x \,|\, \xi \geq y\} = \frac{\mathsf{P}\{\zeta < x, \xi \geq y\}}{\mathsf{P}\{\xi \geq y\}} = \frac{L(x) - F(x,y)}{1 - B(y)}.$$

Corollary. *LST of the function* $E_y(x)$ *has the form:*

$$e_y(s) = [1 - B(y)]^{-1}R(s,y),$$

where $R(s,y) = \int_{x=0}^\infty e^{-sx} \int_{u=y}^\infty dF(x,u)$.

3   Method of Additional Event and Its Modification

Firstly, we present short the classical method of additional event that is very rarely used in papers written in English. This method was introduced by G. P. Klimov (see [4]) and successfully used for analysis of priority queueing systems [3]. Its idea is to give a probability sense to the formal mathematical transforms: LST and generating function (GF). To clarify the idea of the method, we consider two simple examples.

Example 1. Let non-negative RV ξ with DF $A(t)$ is the duration of some random process under consideration. Let, independently of this process behavior, some events (so-called catastrophes) take place and form stationary Poisson arrival process with parameter $q > 0$. Then $a(q) = \int_0^\infty e^{-qt}\mathrm{d}A(t)$ is probability that catastrophes do not appear during duration of the process.

Example 2. Let ξ be a number of customers coming to a system during some fixed time interval and $p_k = \mathsf{P}\{\xi = k\}$. Assume that each customer is of red colour with probability z ($0 \le z \le 1$) and of blue colour with probability $1 - z$, independently of other customers' colours. Then $P(z) = \sum_{k=0}^\infty p_k z^k$ is probability that only red customers come to the system during this interval (or blue customers do not come during it).

By this way, we give the probability sense to LST $a(q)$ and GF $P(z)$ when $q > 0$ and $0 \le z \le 1$, consequently. Now, we can calculate these functions as probabilities of proper events using, if it is necessary, the principle of analytic continuation.

Let e.g. $P(z,t) = \sum_{k=0}^\infty P_k(t)z^k$ be GF of number of customers present in a system at time instant t. Then, for $0 \le z \le 1$, function $P(z,t)$ is probability that there are no blue customers in the system at time instant t. Let $\pi(z,q) = \int_0^\infty e^{-qt}P(z,t)\mathrm{d}t$ be Laplace transform of $P(z,t)$ with respect to t. Then, $q\pi(z,q)$ is probability that the first catastrophe appears when there are no blue customers in the system.

Later on, for analysis of the system presented in Sect. 2, we shall use some modification of the classical method of additional event. We assume that: 1) some events (catastrophes) take place independently of the behavior of a system under consideration, they form Poisson arrival process with parameter q, $q > 0$ (this proposition does not distinguish from proper classical one); 2) an arbitrary customer of volume x has red colour with probability e^{-sx}, $s > 0$ or blue one with probability $1 - e^{-sx}$, independently of other customers' colours (this proposition is the generalization of proper classical one).

Then, the functions introduced in Sect. 2 have the following probability sense [13]: $\varphi(s)$ is probability that an arbitrary customer is red; $\beta(q)$ is probability that there are no catastrophes during arbitrary customer's service; $\alpha(s,q)$ is the joint probability that an arbitrary customer is red and there are no catastrophes during his service; $e_y(s) = [1 - B(y)]^{-1}R(s,y)$ is probability that a customer on service is red, under condition that time y has passed from the beginning of his service; $q\delta(s,q)$ is probability that the first catastrophe takes place in the system when there are no blue customers in it.

Presentation of the method of additional event and its modification is also the aim of this paper. Of course, it is possible to use other methods for analysis of the unreliable system with random volume customers, e.g. this system can be interpreted as a system with vacations (see e.g. [7]), but, in this paper, we demonstrate possibilities of our method that can be used also for analysis of other queueing models. Note that this method in its modification form was used for analysis of the system $M/G/1/\infty$ with random volume customers and preemptive service discipline (see [8]).

As an example, let us consider an application of this method to determine the function $\delta(s,q)$ for the system $M/G/1/\infty$ with reliable server. Note that this queue was investigated earlier by other method (see e.g. [12]). Below, we call 0-moments the moments of service beginning or termination (see [3]).

Assume that a busy period of the system begins at time instant 0 and continues at time t. Let $\Pi(x,y,t)\,dy = \mathsf{P}\{\sigma(t) < x,\ \xi^*(t) \in [y; y+dy)\}$ be probability that the total customers' volume $\sigma(t)$ is less than x at time instant t, and time y has passed from the last 0-moment to the instant t. Let

$$\pi(s,y,q) = \int_{x=0}^{\infty} \int_{t=0}^{\infty} e^{-sx-qt}\, d_x \Pi(x,y,t)\, dt.$$

Then, $q\pi(s,y,q)dy$ is probability that the first catastrophe on the busy period occurs when there are no blue customers in the system and time y has passed from the last 0-moment. Denote by $\pi(s,q) = \int_0^\infty \pi(s,y,q)dy$. Then, $q\pi(s,q)$ is probability that the first catastrophe on a separate busy period occurs when there are no blue customers in the system. Let $\Pi(t)$ be DF of busy period of the system.

Firstly, we determine probability $\pi(s,0,q)$ that there are no blue customers in the system at some epoch of service termination and catastrophes do not appear to this epoch inside of the busy period.

Lemma 2. *The function $\pi(s,0,q)$ is determined by the relation*

$$\pi(s,0,q) = \frac{\varphi(s)[\beta(q+a-a\varphi(s)) - \pi(q)]}{\varphi(s) - \beta(q+a-a\varphi(s))},$$

where $\pi(q) = \int_0^\infty e^{-qt}d\Pi(t)$ is LST of the busy period.

Proof of the lemma follows from the appropriate statement in [3] (p. 18), where z must be substituted by $\varphi(s)$.   □

Lemma 3. *The functions $\pi(s,y,q)$ and $\pi(s,0,q)$ are connected by the following relation:*

$$\pi(s,y,q) = e^{-(q+a-a\varphi(s))y}\left[1 + \frac{\pi(s,0,q)}{\varphi(s)}\right] R(s,y).$$

Proof. The first catastrophe inside of the busy period occurs in the system at time instant when there are no blue customers in it and the time y passed from the last 0-moment, iff

1) either the first catastrophe occurs during the first red customer service when time y has passed from the beginning of his service (probability of this event is $R(s,y)qe^{-qy}dy$), and only red customers arrived during time y (probability of this event is $e^{-a(1-\varphi(s))y}$); therefore the complete probability of this event is $qe^{-(q+a-a\varphi(s))y}R(s,y)dy$;

2) or, at some 0-moment inside of the busy period, there were no blue customers in the system and there were no catastrophes before this moment (probability of this event is $\pi(s, 0, q)$), the customer on service was red after time y has passed from his service beginning (probability of this event is $\{[1 - B(y)]\varphi(s)\}^{-1}R(s, y)$), and a catastrophe appeared at this moment and during time y blue customers did not arrive; the complete probability of this event is

$$\frac{\pi(s, 0, q)}{\varphi(s)}qe^{-(q+a-a\varphi(s))y}R(s, y)dy.$$

By summing obtained probabilities, we obtain the statement of the lemma.   □

If we substitute the function $\pi(s, 0, q)$ from lemma 2 to the relation in the statement of lemma 3, we obtain the following theorem.

Theorem 1. *a)* *The function* $\pi(s, y, q)$ *is determined by the following relation:*

$$\pi(s, y, q) = \gamma(s, y, q)\frac{\varphi(s) - \pi(q)}{\varphi(s) - \beta(q + a - a\varphi(s))}, \tag{1}$$

where

$$\gamma(s, y, q) = e^{-(q+a-a\varphi(s))y}R(s, y). \tag{2}$$

b) *The function* $\pi(s, q)$ *is determined by the relation*

$$\pi(s, q) = \int_0^\infty \pi(s, y, q)\,dy$$

$$= \frac{\varphi(s) - \alpha(s, q + a - a\varphi(s))}{q + a - a\varphi(s)} \cdot \frac{\varphi(s) - \pi(q)}{\varphi(s) - \beta(q + a - a\varphi(s))}.$$

Denote by $P(x, y, t)dy = \mathsf{P}\{\sigma(t) < x, \ \xi^*(t) \in [y; y + dy)\}$ probability that the total customers volume is less than x at the time instant t and time y has passed from the last 0-moment to this instant. Then,

$$q\delta(s, y, q)dy = q\int_{x=0}^\infty \int_{t=0}^\infty e^{-sx-qt}d_xP(x, y, t)\,dy\,dt$$

is the probability that the first catastrophe occurs when there are no blue customers in the system and time y has passed from the last 0-moment; $q\delta(s, q) = q\int_0^\infty \delta(s, y, q)dy$ is the probability that the first catastrophe occurs when there are no blue customers in the system.

Theorem 2. *a)* *The function* $\delta(s, y, q)$ *is determined by the following relation:*

$$\delta(s, y, q) = \frac{e^{-(q+a)y}}{q + a - a\pi(q)}\left[q + a + \frac{ae^{a\varphi(s)y}(\varphi(s) - \pi(q))}{\varphi(s) - \beta(q + a - a\varphi(s))}R(s, y)\right].$$

b) *The function $\delta(s,q)$ is determined by the following relation:*

$$\delta(s,q) = \int_0^\infty \delta(s,y,q)\,dy = [q + a - a\pi(q)]^{-1}$$

$$\times \left\{1 + \frac{\varphi(s) - \alpha(s, q + a - a\varphi(s))}{q + a - a\varphi(s)} \cdot \frac{a[\varphi(s) - \pi(q)]}{\varphi(s) - \beta(q + a - a\varphi(s))}\right\}.$$

Proof. Determine the probability $q\delta(s,y,q)dy$. A proper event takes place iff:

1) either the first catastrophe occurs on the first interval when server is free (probability of this event is $e^{-(q+a)y}$);
2) or the first busy period begins earlier than a catastrophe appears (probability of this event is $a/(q+a)$) and the first catastrophe occurs on this period when there are no blue customers in the system, and time y has passed from the last 0-moment (probability of this event is $q\pi(s,y,q)dy$);
3) or there were no catastrophes during the first interval when the server was free nor during the first busy period (probability of this event is $a\pi(q)/(q+a)$), and further the process behaves as from the start (it is clear that epochs of busy periods terminations are regeneration points of the process $\sigma(t)$).

As a result, we have:

$$q\delta(s,y,q)dy = qe^{-(q+a)y}dy + \frac{aq}{q+a}\pi(s,y,q)dy + \frac{aq\pi(q)}{q+a}\delta(s,y,q)dy,$$

whereas we obtain the first statement of the theorem.   □

The last relation in the second statement coincides with results obtained earlier (see [12]).

4   The Model and Preliminary Results

Consider a system $M/G/1/\infty$ and assume that its server is reliable in busy state. If T is an epoch of service termination when there are no waiting customers and other customers do not arrive to the system during time t, the server can be broken on time interval $[T; T + t)$ with probability $E(t)$. After breakage, the server restores during some random time ψ. Denote by $H(t) = \mathsf{P}\{\psi < t\}$ DF of RV ψ. The volume of a customer ζ and his service time ξ are determined by the joint DF $F(x,t) = \mathsf{P}\{\zeta < x, \xi < t\}$. We assume that service time, the time of reliable state of the server and renewal time are independent RVs. For the considered system, we determine the function

$$\delta(s,q) = \int_0^\infty e^{-qt}\left[\int_{x=0}^\infty e^{-sx}d_x D(x,t)\right]dt,$$

where $D(x,t) = \mathsf{P}\{\sigma(t) < x\}$ is DF of total volume at time instant t.

Denote by $\pi(q)$ LST of busy period of the system under consideration when $E(t) \equiv 0$ (this is the busy period of the reliable system $M/G/1/\infty$). Let $\pi_n(q) = [\pi(q)]^n$ be LST of DF of so-called n-period (busy period that begins from the moment when there are n customers in the system, $n = 1, 2, \ldots$). Let $\Pi^{(n)}(x, y, t)dy$ be probability that the total customers volume $\sigma(t)$ is less than x at time instant t, and time y has passed from the last 0-moment, under assumption that this n-period does not terminate at time instant t.

Introduce the notations:

$$\pi_n(s, y, q) = \int_{x=0}^{\infty} \int_0^{\infty} e^{-sx-qt} d_x \Pi^{(n)}(x, y, t) dt,$$

$$\pi_n(s, q) = \int_0^{\infty} \pi_n(s, y, q) dy.$$

Then, $q\pi_n(s, y, q)dy$ is probability that the first catastrophe on n-period occurs when there are no blue customers in the system and time y has passed from the last 0-moment, $q\pi(s, y, q)dy$ is probability of an analogous event for the reliable system.

Theorem 3. *The following relations take place:*

$$\pi_n(s, y, q) = \frac{(\varphi(s))^n - (\pi(q))^n}{\varphi(s) - \beta(q + a - a\varphi(s))} e^{-(q+a-a\varphi(s))y} R(s, y),$$

$$\pi_n(s, q) = \frac{(\varphi(s))^n - (\pi(q))^n}{\varphi(s) - \beta(q + a - a\varphi(s))} \cdot \frac{\varphi(s) - \alpha(s, q + a - a\varphi(s))}{q + a - a\varphi(s)}. \tag{3}$$

Proof. As it follows from relations (1) and (2), the relation (3) can be presented as:

$$\pi_n(s, y, q) = \frac{(\varphi(s))^n - (\pi(q))^n}{\varphi(s) - \pi(q)} \pi(s, y, q), \quad n \geq 1,$$

or, if we treat the fraction as a sum of n initial items of geometrical progression with the first item $(\varphi(s))^{n-1}$ and denominator $\pi(q)/\varphi(s)$,

$$q\pi_n(s, y, q)dy = (\varphi(s))^{n-1} q\pi(s, y, q)dy + (\varphi(s))^{n-2}\pi(q)q\pi(s, y, q)dy$$
$$+ \cdots + \varphi(s)(\pi(q))^{n-2} q\pi(s, y, q)dy + (\pi(q))^{n-1} q\pi(s, y, q)dy. \tag{4}$$

Relation (4) can be proved by the modified method of additional event.

Assume that the first catastrophe inside of n-period occurs when all customers present in the system are red and time y has passed from the last 0-moment. This event takes place iff:

a) either the first catastrophe occurs inside of busy period connected with the first (from n) served customer, there are no blue customers in the system at this time instant, time y has passed from the last 0-moment (probability of this event is $q\pi(s, y, q)dy$) and other $n-1$ customers present in the system at the moment of beginning of the n-period were red (probability of this event is $(\varphi(s))^{n-1}$);

b) or there were no catastrophes inside of busy period connected with the first served customer (probability of this event is $\pi(q)$), but the first catastrophe occurs inside of busy period connected with the second served customer (from presented ones at the beginning of the n-period) at time instant when all customers present in the system are red, time y has passed from the last 0-moment (probability of this event is $\pi(q)q\pi(s,y,q)dy$) and other $n-2$ customers from those presented at the beginning of the n-period were red (probability of this event is $(\varphi(s))^{n-2}$);

c) or, finally, there were no catastrophes during initial $n-1$ busy periods and the first catastrophe occurs inside of the last one at time instant when all customers present in the system are red and time y has passed from the last 0-moment (probability of this event is $(\pi(q))^{n-1}q\pi(s,y,q)dy$).

Summing obtained probabilities, we obtain the relation (4). The rest of the proof is evident.   □

5   Main Statements

Denote by $\varepsilon(q)$ and $h(q)$ LSTs of DFs $E(t)$ and $H(t)$, respectively. As a regeneration period of the system we mean the time interval between neighbouring epochs when the system becomes empty after service termination. It is follows from [3, p. 46] that, for the system under consideration, LST $r(q)$ of DF of a regeneration period is determined by the following relation:

$$r(q) = \frac{a}{q+a}[1 - \varepsilon(q+a)]\pi(q) + \varepsilon(q+a)h(q+a-a\pi(q)). \tag{5}$$

Denote by $\omega(t)$ the time that has passed from the beginning of a regeneration period to some time moment t inside it.

Let $P(x,y,t)dy = \mathsf{P}\{\sigma(t) < x, \omega(t) \in [y; y+dy)\}$. Then

$$\overline{\delta}(s,y,q) = \int_0^\infty e^{-qt}\left[\int_{x=0}^\infty e^{-sx}d_x D(x,t\,|\,\omega(t) = y)\right]dt$$

be the Laplace transform (with respect to t) of LST (with respect to x) of DF of total volume of customers present in the system t time units after beginning of a regeneration period, if time y has passed from the last 0-moment inside of the period.

Theorem 4. *The function $\overline{\delta}(s,y,q)$ is determined by the following relation:*

$$\overline{\delta}(s,y,q) = [1 - E(y)]e^{-(q+a)y} + \frac{\varepsilon(q+a)}{q+a}e^{-(q+a-a\varphi(s))y}$$

$$\times \frac{\varphi(s) - \pi(q)}{\varphi(s) - \beta(q+a-a\varphi(s))}R(s,y)$$

$$+ \frac{\varepsilon(q+a)\left[h(q+a-a\varphi(s)) - h(q+a-a\pi(q))\right]}{\varphi(s) - \beta(q+a-a\varphi(s))}e^{-(q+a-a\varphi(s))y}R(s,y)$$

$$+ \varepsilon(q+a)[1 - H(y)]e^{-(q+a-a\varphi(s))y}. \tag{6}$$

Proof. A time interval when the server is busy that begins either from beginning of service of an arriving customer, or from the epoch of server breakage, and terminates at the nearest time moment when the server is in good repair and there are no customers in the system, we call generalized busy period. Therefore, it is a time interval of two possible types [3]: 1) a generalized busy period begins from customer service; 2) it begins from server breakage. Probabilities that the regeneration period involves generalized busy periods of types 1 and 2 equal $a(q + a)^{-1}[1 - \varepsilon(q + a)]$ and $\varepsilon(q + a)$, respectively.

Recall that $q\pi(s, y, q)dy$ is probability that the first catastrophe on the busy period of the reliable system $M/G/1/\infty$ occurs when there are no blue customers in the system and time y has passed from the last 0-moment. It is clear that, for the system under consideration, probability of an analogous event on the generalized busy period of the first type equals $q\pi_n(s, y, q)dy$, i.e. the distribution of the total customers volume on this period is determined by the function $\pi(s, y, q)$.

Consider a generalized busy period of the second type. It is clear that blue customers form Poisson arrival process with parameter $a(1 - \varphi(s))$. Hence, $e^{-(q+a-a\varphi(s))y}$ is probability that, during time y, catastrophes do not occur and blue customers do not arrive; $q[1 - H(y)]e^{-(q+a-a\varphi(s))y}dy$ is probability that, during renewal period having duration greater than y, the first catastrophe occurs after time y from the beginning of the period, and only red customers arrive to the system before the catastrophe. Probability that catastrophes do not occur and n customers arrive to the system during the renewal period is equal to $\int_0^\infty \frac{(au)^n}{n!}e^{-(q+a)u}dH(u)$. Then, probability $qG(s, y, q)dy$ that period of the second type involves service of customers and the first catastrophe inside it occurs when there no blue customers in the system, and time y has passed from the last 0-moment is equal to

$$qG(s, y, q)dy = \sum_{n=1}^{\infty} \int_{u=0}^{\infty} \frac{(au)^n}{n!} e^{-(q+a)u} q\pi_n(s, y, q)dy\, dH(u),$$

where function $\pi_n(s, y, q)$ is determined by relation (3), whereas we obtain after some transformations:

$$qG(s, y, q)dy = qe^{-(q+a-a\varphi(s))y}dy \frac{h(q + a - a\varphi(s)) - h(q + a - a\pi(q))}{\varphi(s) - \beta(q + a - a\varphi(s))} R(s, y).$$

Now, we can obtain the relation (6) using (5) and formula of total probability.   □

Denote by $\delta(s, y, q)$ Laplace transform with respect to t of total customers volume $\sigma(t)$ under condition that time y has passed from the last 0-moment.

The next statement follows from Theorem 4 and can be proved analogously as Theorem 2.

Theorem 5. *a) Function $\delta(s,y,q)$ is determined by the following relation:*

$$\delta(s,y,q) = \left\{1 - \frac{a}{q+a}[1 - \varepsilon(q+a)]\pi(q) - \varepsilon(q+a)h(q+a-a\pi(q))\right\}^{-1}$$

$$\times \left\{[1 - E(y)]e^{-(q+a)y} + \frac{ae^{-(q+a-a\varphi(s))y}}{q+a}[1 - \varepsilon(q+a)]\frac{R(s,y)(\varphi(s) - \pi(q))}{\varphi(s) - \beta(q+a-a\varphi(s))}\right.$$

$$+ \varepsilon(q+a)[1 - H(y)]e^{-(q+a-a\varphi(s))y}$$

$$+ \varepsilon(q+a)e^{-(q+a-a\varphi(s))y}\frac{h(q+a-a\varphi(s)) - h(q+a-a\pi(q))}{\varphi(s) - \beta(q+a-a\varphi(s))}R(s,y)\left.\right\}.$$

b) Function $\delta(s,q) = \int_0^\infty e^{-qt}\mathsf{E}e^{-s\sigma(t)}\,dt$ is determined by the relation:

$$\delta(s,q) = \int_0^\infty \delta(s,y,q)\,dy$$

$$= \left\{1 - \frac{a}{q+a}[1 - \varepsilon(q+a)]\pi(q) - \varepsilon(q+a)h(q+a-a\pi(q))\right\}^{-1}\left\{\frac{1 - \varepsilon(q+a)}{q+a}\right.$$

$$+ \frac{a[1 - \varepsilon(q+a)][\varphi(s) - \alpha(s,q+a-a\varphi(s))][\varphi(s) - \pi(q)]}{(q+a)(q+a-a\varphi(s))[\varphi(s) - \beta(q+a-a\varphi(s))]} \tag{7}$$

$$+ \frac{\varepsilon(q+a)[\varphi(s) - \alpha(s,q+a-a\varphi(s))]}{q+a-a\varphi(s)} \cdot \frac{h(q+a-a\varphi(s)) - h(q+a-a\pi(q))}{\varphi(s) - \beta(q+a-a\varphi(s))}$$

$$+ \left.\frac{\varepsilon(q+a)[1 - h(q+a-a\varphi(s))]}{q+a-a\varphi(s)}\right\}.$$

Corollary. *If $\rho = a\beta_1 < 1$, a steady state exists for the system under consideration, i.e. $\sigma(t) \Rightarrow \sigma$ in the sense of a weak convergence. LST $\delta(s)$ of DF $D(x) = \lim_{t\to\infty} D(x,t) = \mathsf{P}\{\sigma < x\}$ of RV σ is determined by the following relation:*

$$\delta(s) = \lim_{q\to 0} q\delta(s,q) = \frac{1-\rho}{1 - \varepsilon(a)(1 - ah_1)}\left\{\left[1 - \varepsilon(a) + \frac{\varepsilon(a)(1 - h(a - a\varphi(s)))}{1 - \varphi(s)}\right]\right.$$

$$\times \left.\left[1 + \frac{\varphi(s) - \alpha(s,a - a\varphi(s))}{\beta(a - a\varphi(s)) - \varphi(s)}\right]\right\}, \tag{8}$$

where h_1 is the first moment of DF $H(t)$.

Using appropriate relations from Sect. 2, we can calculate moments of the total volume or their Laplace transforms. Note that, if $\varepsilon(q) \equiv 0$, we obtain the known relations for the reliable system $M/G/1/\infty$ (see [13]).

For example, Laplace transform $\delta_1(q) = \int_0^\infty e^{-qt}\delta_1(t)dt$ of the mean total customers volume $\delta_1(t) = \mathsf{E}\sigma(t)$ has the following form:

$$\delta_1(q) = \left\{ 1 - \frac{a}{q+a}[1 - \varepsilon(q+a)]\pi(q) - \varepsilon(q+a)h(q+a-a\pi(q)) \right\}^{-1}$$

$$\times \left[\varepsilon(q+a)(h(q) - h(q+a-a\pi(q))) \right.$$

$$+ \frac{a(1 - \varepsilon(q+a))(1 - \pi(q))}{q+a} \left] \frac{\Delta_s(1)\alpha(s,q)|_{s=0}}{q(\beta(q)-1)} \right.$$

$$+ \frac{a\varphi_1}{q^2} \left[\varepsilon(q+a)(1 - h(q+a-a\pi(q))) + \frac{(1 - \varepsilon(q+a))(q+a-a\pi(q))}{q+a} \right] \right\},$$

whereas we obtain the first and second moments of RV σ in the following form:

$$\delta_1 = \mathsf{E}\sigma = a\alpha_{11} + \frac{a^2\beta_2\varphi_1}{2(1-\rho)} + \frac{a^2 h_2\varepsilon(a)\varphi_1}{2(1 - \varepsilon(a) + a\varepsilon(a)h_1)}, \tag{9}$$

$$\delta_2 = \mathsf{E}\sigma^2 = a(\alpha_{21} + a\varphi_1\alpha_{12}) + \frac{a^3\beta_2\varphi_1\alpha_{11}}{1-\rho} + \frac{a^2\beta_2\varphi_2}{2(1-\rho)} + \frac{a^3\beta_3\varphi_1^2}{3(1-\rho)}$$

$$+ \frac{a^4\beta_2^2\varphi_1^2}{2(1-\rho)^2} + \frac{a^3\varepsilon(a)h_2\varphi_1\alpha_{11}}{1 - \varepsilon(a) + a\varepsilon(a)h_1} + \frac{a^2\varepsilon(a)h_2\varphi_2}{2(1 - \varepsilon(a) + a\varepsilon(a)h_1)} \tag{10}$$

$$+ \frac{a^3\varepsilon(a)h_3\varphi_1^2}{3(1 - \varepsilon(a) + a\varepsilon(a)h_1)} + \frac{a^4\varepsilon(a)h_2\varphi_1^2}{2(1-\rho)(1 - \varepsilon(a) + a\varepsilon(a)h_1)}.$$

6 The Case of the System Unreliable Also When Server Is Busy

Our model can be generalized to the case of unreliable server also when it is busy. Let, in addition, the server can be broken on time interval $[T; T+t)$, where T is a moment of service beginning, with probability $G(t)$ (we assume that service of the customer does not terminate before time instant $T+t$). If this event takes place, the service of the customer is interrupted and will be continued after server's renewal. Denote by $X(t)$ DF of the renewal period. Let $g(q)$ and $\chi(q)$ be LSTs of the functions $G(t)$ and $X(t)$, respectively, and denote by χ_i, $i = 1, 2, \ldots$, the ith moment of the renewal time.

Obviously, in this case, the problem of determination of total customers volume distribution comes to previous one solved in Sect. 5, if service time of a customer is substituted by the time from beginning to termination of his service. This time is called the time of customer presence on server. Let $\kappa(q)$ be LST of DF of this time (taking into consideration possible breakages and renewals). Then, when $\varepsilon(q) \equiv 0$, instead of the equation $\pi(q) = \beta(q+a-a\pi(q))$, we have the following functional equation for busy period of the system under consideration: $\pi(q) = \kappa(q+a-a\pi(q))$.

Denote by $P(z,t)$ the generating function of number of server's breakages during customer service time t, under assumption that the total service time $y \geq t$. Let $B(t \mid \zeta = x) = \mathsf{P}\{\xi < t \mid \zeta = x\}$ is conditional DF of service time of a customer, under condition that his volume equals x, $\kappa(q \mid \zeta = x)$ is LST of DF of the time of customer presence on server, if his volume equals x. It is clear that

$$\kappa(q \mid \zeta = x) = \int_0^\infty e^{-qt} P(\chi(q), t) \, dB(t \mid \zeta = x), \tag{11}$$

where the function $P(z,t)$ can be determined via its Laplace transform $p(z,q)$ (see e.g. [13]):

$$p(z,q) = \int_0^\infty e^{-qt} P(z,t) \, dt = \frac{1 - g(q)}{q[1 - zg(q)]}.$$

Let ω be the time of customer presence on server, $\Gamma(x,t) = \mathsf{P}\{\zeta < x, \omega < t\}$ be joint DF of customer volume ζ and RV ω. We denote by $\gamma(s,q)$ double LST of the function $\Gamma(x,t)$:

$$\gamma(s,q) = \int_{x=0}^\infty \int_{t=0}^\infty e^{-sx-qt} \, d\Gamma(x,t).$$

In particular, $\varphi(s) = \gamma(s,0)$, $\kappa(q) = \gamma(0,q)$. It follows from (11) that

$$\gamma(s,q) = \int_{x=0}^\infty e^{-sx} \kappa(q \mid \zeta = x) \, dL(x) = \int_{x=0}^\infty \int_{t=0}^\infty e^{-sx-qt} P(\chi(q), t) \, dF(x,t). \tag{12}$$

So, the problem of determination of process $\sigma(t)$ characteristics comes to analogous problem for the system $M/G/1/\infty$ with unreliable server in free state only, if we assume that the joint DF of customer volume and his service time is $\Gamma(x,t)$. Let κ_i be the ith moment of RV ω and γ_{ij} be the mixed $(i+j)$th moment of the random vector (ζ, ω). Then, for the system under consideration, the function $\delta(s,q)$ is determined by the relation (7), where we have to replace $\beta(q)$ by $\kappa(q)$ and $\alpha(s,q)$ by $\gamma(s,q)$.

It is clear that steady state exists for the system under consideration, if the inequality $\rho^* = a\kappa_1 < 1$ holds. The function $\delta(s)$ can be determined by relation (8) with the same previously made replacements.

7   Special Cases and Numerical Results

In this section we analyze some special cases of investigated in Sect. 6 model. Assume additionally that $G(t) = 1 - e^{-dt}$, $d > 0$. In this case, we obtain $P(z,t) = e^{-(1-z)dt}$, and it follows from (12) that

$$\gamma(s,q) = \int_{x=0}^\infty \int_{t=0}^\infty e^{-sx-(q+d-d\chi(q))t} dF(x,t) = \alpha(s, q+d - d\chi(q)), \tag{13}$$

whereas we obtain:

$$\kappa(q) = \beta(q + d - d\chi(q)).\qquad(14)$$

Then, for function $\delta(s)$ determination we can use relation (8), where $\alpha(s, q)$ is substituted by $\alpha(s, q + d - d\chi(q))$ and $\beta(q)$ by $\beta(q + d - d\chi(q))$.

The moments δ_1, δ_2 can be calculated by relations (9), (10), where α_{ij} is substituted by γ_{ij}, and β_i – by κ_i. We can easily obtain that

$$\gamma_{11} = (1 + d\chi_1)\alpha_{11}, \ \gamma_{21} = (1 + d\chi_1)\alpha_{21}, \ \gamma_{12} = (1 + d\chi_1)^2\alpha_{12} + d\chi_2\alpha_{11}, \quad(15)$$

$$\kappa_1 = (1 + d\chi_1)\beta_1, \ \kappa_2 = (1 + d\chi_1)^2\beta_2 + d\chi_2\beta_1,$$
$$\kappa_3 = (1 + d\chi_1)^3\beta_3 + 3d(1 + d\chi_1)\chi_2\beta_2 + d\chi_3\beta_1.\qquad(16)$$

Assume additionally that $E(t) = G(t) = 1 - e^{-dt}$ (probabilities of servers's breakage in free and busy state are determined exponentially with the same parameter d) and customer's service time ξ and volume ζ are connected by the relation $\xi = c\zeta$, $c > 0$. Then, we obtain $\alpha(s, q) = \varphi(s + cq)$ and $\beta(q) = \varphi(cq)$ (see e.g. [12,13]), and mixed moments α_{ij} are determined as $\alpha_{ij} = c^j\varphi_{i+j}$, moments β_i are determined as $\beta_i = c^i\varphi_i$, $i, j = 1, 2, \dots$. In addition, we suppose that renewal periods distributions in server's free or busy state are also the same which means that $H(t) = X(t)$ (so $\chi(q) = h(q)$).

In this case, the relation (8) (taking into consideration replacements $\alpha(s, q)$ by $\gamma(s, q)$ and $\beta(q)$ by $\kappa(q)$) has the form

$$\delta(s) = \left[\frac{1}{1 + dh_1} - ac\varphi_1\right]\left\{\left[1 + \frac{d(1 - h(a\psi(s)))}{a\psi(s)}\right]\right.$$
$$\times \left.\left[1 + \frac{\varphi(s) - \varphi(s + ca\psi(s)) + cd - cdh(a\psi(s)))}{\varphi(ca\psi(s) + cd - cdh(a\psi(s))) - \varphi(s)}\right]\right\},$$

where $\psi(s) = 1 - \varphi(s)$. Initial moments of the total volume can be calculated by formulae (9), (10), where $\alpha_{ij} = \gamma_{ij}$, $\beta_i = \kappa_i$, and the values γ_{ij}, κ_i are calculated by formulae (15), (16).

If we additionally assume that $H(t) = 1 - e^{-rt}$, $L(x) = 1 - e^{-fx}$, then $h_i = i!/r^i$, $\varphi_i = i!/f^i$, $\gamma_{11} = 2c(d + r)/(rf^2)$, $\gamma_{21} = 6c(d + r)/(rf^3)$, $\gamma_{12} = 2c[3c(d + r)^2 + 2df]/(r^2f^3)$, $\kappa_1 = c(d + r)/(rf)$, $\kappa_2 = 2c\left[c(d + r)^2 + df\right]/(rf)^2$, $\kappa_3 = 6c\left[c^2(d + r)^3 + 2cd(d + r)f + df^2\right]/(rf)^3$.

For example, in this case, the relations for the first moments of steady-state total volume have the form:

$$\delta_1 = \frac{a}{rf}\left\{\frac{d}{d+r} + \frac{c(d+r)[2rf - ca(d+r)] + acdf}{f[rf - ca(d+r)]}\right\}.$$

$$
\begin{aligned}
\delta_2 &= \frac{d(-2ad + 2ar - 2dr + a^2dr - 2r^2 + a^2r^2)}{c(d+r)^3(-acd - acr + fr)} + \frac{2(2a^2cd + acdr + acr^2)}{f^3r^2} \\
&+ \frac{2(a^2c^2d^2 + 2a^2c^2dr + a^2c^2r^2)}{f^4r^2} + \frac{2ad^2 - 2adr + 2d^2r - a^2d^2r + 2dr^2 - a^2dr^2}{cfr(d+r)^3} \\
&+ \frac{2(a^2d^2 - d^2r^2 - 2dr^3 - r^4)}{f^2r^2(d+r)^2} + \frac{2(a^2d^2 + 2ad^2r + 2adr^2 + d^2r^2 + 2dr^3 + r^4)}{(d+r)^2(-acd - acr + fr)^2}.
\end{aligned}
\tag{17}
$$

Formulae (17) are important not only from the theoretical point of view. As it was discussed in [13] (p. 262–266), these characteristics are also used in approximation of loss characteristics for analogous (to the mentioned above) queueing system but with limited total volume. Assume now that we analyze single–server queueing model with non–homogeneous customers, unreliable server and limited (by value V) total volume (which means that $\sigma(t) \le V$). Then we introduce, for example, characteristic called loss probability, which is usually determined by the following relation:

$$P_{loss} = 1 - \int_0^V D_V(V - x)\, dL(x), \tag{18}$$

where $D_V(x)$ is the distribution function of the total customers' volume for this system and $L(x)$ – distribution function of the customer's volume. For the systems with limited memory, where service time of a customer and his volume are dependent, it is often impossible to determine function $D_V(x)$. Then we calculate estimators P_{loss}^* of P_{loss} substituting in (18) distribution $D(x)$ instead of $D_V(x)$ ($D(x)$ is analogous function for the system with unlimited total volume). Moreover, even in the case when total volume is unlimited, we rarely obtain relation for $D(x)$ in exact form. We usually obtain its LST and, on the base of its properties, we can calculate its first two moments δ_1, δ_2 that let us approximate convolution $\Phi(x) = \int_0^x D(V - u)\, dL(u)$ (that is present in (18)) by the function $\Phi^*(x) = \frac{\gamma(q,cx)}{\Gamma(q)}$, where $q = f_1^2/(f_2 - f_1^2), c = f_1/(f_2 - f_1^2)$, $f_1 = \delta_1 + \varphi_1$ and $f_2 = \delta_2 + \varphi_2 + 2\delta_1\varphi_1$. Finally, we use formula: $P_{loss}^* = 1 - \Phi^*(V)$. In Table 1, we present numerical computations for the model with limited total volume. Its characteristics are the same as characteristics of the model discussed at the beginning of this section, but this time total volume is limited by V. We present results for the following values: $a = 1, c = 1, d = 0,5, r = 1, f = 2$ (then $\rho = 0,75$); $a = 1, c = 1, d = 0,25, r = 1, f = 2$ ($\rho = 0,625$) and $a = 0,5, c = 1, d = 0,1, r = 1, f = 2$ ($\rho = 0,275$). As we can see, loss probability also depends strictly on the value of the parameter d which determines how often breakages in the system appear.

Note that our approach of P_{loss} estimation guarantees correct determination of buffer capacity, i.e. the determination of such V that this probability does not exceed a given value.

Table 1. Numerical values of P_{loss} for the system with limited memory and unreliable server

V	$P_{loss}(\rho = 0,75)$	$P_{loss}(\rho = 0,625)$	$P_{loss}(\rho = 0,275)$
1	0,7416	0,5980	0,3077
2	0,5330	0,3530	0,1042
3	0,3796	0,2077	0,0360
4	0,2690	0,1220	0,0125
5	0,1900	0,0716	0,0044
6	0,1339	0,0420	0,0015
7	0,0942	0,0246	0,0005
8	0,0662	0,0144	0,0002
9	0,0465	0,0084	0,0001
10	0,0326	0,0049	$2,4 \cdot 10^{-5}$

8 Conclusion and Final Remarks

In the paper, we present the modified method of additional event that is very rarely used in English scientific literature and use this method to investigate single-server queueing system with non-homogeneous customers, unreliable server and unlimited total volume. For the analyzed model, we obtain characteristics of the total volume both in stationary and nonstationary mode. We also calculate first two moments of the steady-state total volume and show loss probability estimators calculations for the analogous model with limited memory that gives us the possibility to determine buffer space capacity of the node of computer or communication network. Investigations show possibility of using method of additional event in the case of complicated models. In addition, we prove that the character of dependency between customer's volume and his service time has influence on characteristics of total volume and estimators of loss characteristics.

References

1. Aleksandrov, A.M., Katz, B.A.: Non-homogeneous demands arrival servicing. Izv. AN SSSR. Tekhnicheskaya Kibernetika. **2**, 47–53 (1973). (in Russian)
2. Bocharov, P.P., D'Apice, C., Pechinkin, A.V., Salerno, S.: Queueing Theory. VSP, Utrecht-Boston (2004)
3. Gnedenko, B.V., Danielyan, E.A., Dimitrov, B.N., Klimov, G.P., Matveev, V.F.: Priority Queues. Moscow University Edition, Moscow (1973). (In Russian)
4. Klimov, G.P.: Stochastic Service System. Nauka, Moscow (1966). (In Russian)
5. König, D., Stoyan, D.: Methoden der Bedienungstheorie. Akademie-Verlag, Berlin (1976)
6. Sengupta, B.: The spatial requirements of M/G/1 queue or: how to design for buffer space. Lect. Notes Contr. Inf. Sci. **60**, 547–562 (1984). https://doi.org/10.1007/BFb0005191

7. Tian, N., Zhang, Z.G.: Vacation Queueing Models Theory and Applications. ISORMS, vol. 93. Springer, Boston (2006). https://doi.org/10.1007/978-0-387-33723-4

8. Tikhonenko, O.M.: Determination of the characteristics of the total amount of customers in queues with absolute priority. Automation and Remote Control. vol. 60, part 2, no. 8, p. 1205–1210 (1999)

9. Tikhonenko, O.M.: Distribution of total message flow in single-server queueing system with group arrival. Autom. Remote Control **46**(11), 1412–1416 (1985)

10. Tikhonenko, O.M.: The problem of determination of the summarized messages volume in queueing system and its applications. J. Inf. Process. Cybern. **32**(7), 339–352 (1987)

11. Tikhonenko, O.M.: Distribution of total flow of calls in group arrival queueing systems. Autom. Remote Control **48**(11), 1500–1507 (1987)

12. Tikhonenko, O.: Queueing Models in Computer Systems. Universitetskoe, Minsk (1990). (In Russian)

13. Tikhonenko, O.M.: Computer Systems Probability Analysis. Akademicka Oficyna Wydawnicza EXIT, Warsaw (2006). (In Polish)

14. Tikhonenko, O., Ziółkowski, M.: Single-server queueing system with external and internal customers. Bull. Polish Acad. Sci. Tech. Sci. **66**(4), 539–551 (2018). https://doi.org/10.24425/124270

Infinite-Server Queue Model $MMAP_k(t)|G_k|\infty$ with Time Varying Marked Map Arrivals of Customers and Occurrence of Catastrophes

Ruben Kerobyan[1(✉)], Khanik Kerobyan[2], Carol Shubin[2], and Phu Nguyen[2]

[1] University of California San Diego, San Diego, CA, USA
rkeroby@ucsd.edu
[2] California State University Northridge, Northridge, CA, USA
khanik.kerobyan@csun.edu

Abstract. In the present paper, the infinite-server queue model $MMAP_k(t)|G_k|\infty$ in transient MMAP random environment with time varying marked MAP arrival of k types of customers subject to catastrophes is considered. The transient joint probability generating functions (PGF) of the number of different types of customers present in the model at moment t and the number of different types of customers departing from the system in the time interval $(0, t]$ are found. The Laplace-Stieltjes transform (LST) of total volume of customers being in service at moment t is defined. The basic differential equations for joint probability generating functions of the number of busy servers and served customers for transient and stationary random environment are obtained.

Keywords: Marked MAP · Infinite-server queue · Catastrophes · MMAP random environment

1 Introduction

Infinite-server (IS) queuing models are widely used for modelling and performance evaluation in different fields such as wired and wireless computer and telecommunication systems and networks, inventory and reliability systems, biology and physics, banking and insurance systems. As shown by the large number of measurements the traffic in modern systems and networks has a heterogeneous, non-stationary, bursty nature and is characterized by self-similarity, long range dependence, and correlation [1]. To describe this type of traffic, finite Markov processes and their generalizations are often used [2,3]: Markov Arrival Process (MAP), Batch MAP (BMAP), Marked MAP (MMAP), Markov additive processes. For more details of MMAP and MAP traffic properties and their applications see [4,9] and references therein.

The infinite-server (IS) queues $M(t)|G|\infty$ with time varying arrival rate of customers and general distributed service times have been considered [10,19] and references therein.

© Springer Nature Switzerland AG 2020
P. Gaj et al. (Eds.): CN 2020, CCIS 1231, pp. 171–184, 2020.
https://doi.org/10.1007/978-3-030-50719-0_13

The IS queue model $BM(t)|G|\infty$ with time-varying batch arrival has been studied in [10]. By using the supplementary variable method, the Laplace transforms (LT) of joint probability generating function (PGF) of queue size, busy period, number of customers being served at moment t, and number of served customers in time interval $(0, t]$ are obtained. In [11], these results are generalized for the $BM(t)|G(t)|\infty$ model with non-homogeneous Poisson arrival of batches of customers, when both the batch size and service time distributions might depend on arrival time. By using probabilistic methods, they derived PGF of number of busy servers at moment t and PGF of number of served customers in the time interval $(0, t]$. The joint generating function of number of busy servers and number of served customers for IS models $BM(t)|G|\infty$ and $BM(t)|M(t)|\infty$ with time-varying batch arrivals are studied in [12] by using the conditional expectations method. For the IS models $BM(t)|G|\infty$ and $BM(t)|G|\infty$ with time-varying arrivals in [13,15] the physics and main properties of the models, PGFs of queue size, departure processes, correlation functions, stationary and transient queue size distributions were studied. In [16] the model $BM(t)|G|\infty$ with time-varying arrival of batches is considered. The customers in each group may be of different types and may have dependent service times. The PGF of the transient distribution of the numbers of customers of various types present in the system, and PGF of the corresponding departure counting process is derived. The IS models $BMAP|G|\infty$ and $BMAP(t)|G|\infty$ with homogeneous and inhomogeneous batch MAP arrival of customers are studied in [8]. The PGF of the number of busy servers and its moments are found by using matrix analytic methods. In [17], the basic differential equations, and their solutions for PGF of queue size and PGF of number of served customers in the time interval $(0, t]$ for the IS models $M(t)|G|\infty$ and $BM(t)|G|\infty$ with homogeneous and time-varying arrivals are derived. The IS model $MMAP_k(t)|G_k|\infty$ with nonhomogeneous marked MAP arrival of customers is studied in [18,19]. The basic differential equations for PGF of queue size of different types of customers and their matrix-exponential solutions are obtained. The infinite-server queue models $G(t)|G^D|\infty$ and $BG(t)|G^D|\infty$ having general arrival process time-varying rate and weakly dependent service times of customers (they established heavy-traffic limit for queue length process) are considered in [20]. A Gaussian approximation for the queue size distribution is found.

Infinite-server queue models with non-Markovian arrival and general service time were considered by number of authors, for example [8,21,25] and references therein. The model $PH|G|\infty$ with phase type arrival is studied in [21,23]. By using the first jump method an integral equation for generating function of queue size is found in [23]. By means of conditional expectations method a basic differential equation for generating function of queue size for queue models $PH|G|\infty$ and $N|G|\infty$ are derived in [21,22]. The model $MAP_k|G_k|\infty$ with the structured batch arrival of k types of customers is studied in [26]. By using the supplementary variable method and the dynamic screening method the differential equations for the characteristic function of queue length for IS models $MAP|G|\infty$, $G|G|\infty$, $G_k|G_k|\infty$ are obtained in [24,25,27,28]. The parameters of Gaussian approximations for probability distributions of queue length are found.

The infinite-server queue models in the random environment are studied in [29,33] and references therein. The distribution of queue size of infinite-server models in a semi-Markov (SM) environment has been studied in [29,30,33].

The queue $M|G|\infty$ in a random environment with clearing mechanism is studied in [31,32]. The environmental clearing process is modeled by an m-state irreducible semi-Markov process. The transient and steady-state queue length distributions are obtained by using renewal arguments. The queue models $MAP_k|G_k|\infty$ and $MMAP_k|G_k|\infty$ in an SM environment and catastrophes are studied in [33,34]. By using renewal arguments and the collective mark method the basic differential and integral equations for PGFs of joint distributions of queue size and number of served customers and their solutions are found. The infinite-server queue $MMAP_k(t)|G_k|\infty$ with Poisson stream of catastrophes and nonhomogeneous marked MAP arrival of customers is studied in [18]. The basic differential equations for PGF of queue length of different types of customers are obtained.

In many applications of queuing models such as computer and communication networks, production systems, transportation systems, and banking and insurance systems, the customers can be characterized by a vector of requesting resources. The components of that vector can be deterministic or random quantities. One component, for instance, can describe the number of servers necessary to serve the customer, a second can describe the amount of time necessary to serve the customer, and a third the volume of space necessary to serve the customer and so on. The customers, also, can have some features necessary to be accepted by system. Some components of resource vectors can be discrete (e.g. number of servers, amount of parts) and others can be continuous (e.g. space-volume to save, amount of finances, power).

Despite the importance of resource models of customers in queuing theory, there are very few works devoted to the research of such kind of models, see for example [33–40] and references therein. The IS queue $M|G|\infty$ with random volume of customers was first introduced in [35]. Different modifications of this model were considered in [36,40]. The queue models $MAP_k|G_k|\infty$ and $MMAP_k|G_k|\infty$ with resource vector in random environment are considered in [33,34]. The basic differential equations for joint PGF of different types of customers and accumulated resource vector are found by using the collective marks method.

In this paper we consider some generalizations of results from [18,19,35]. We study the infinite-server queue $MMAP_k(t)|G_k|\infty$ with transient marked MAP random environment and nonhomogeneous, time varying marked MAP arrival of customers and catastrophes. The joint PGF of number of busy servers by different types of customers and numbers of different types of served customers in interval of time $(0,t]$ are studied. The LST of total volume of customers being in service at moment t is defined. The basic differential equations for joint PGF of queue sizes of different types of customers being served at moment t and number of different types of customers served in interval $(0,t]$ are found as well.

2   Model Description

The model consists of infinite number of identical servers; the batches of types of customers arrive according to a non-homogeneous marked Markovian arrival process (MMAP) $\xi_0(t)$. The service of an arriving customer starts immediately. Service times of i th type customers are independent and identically distributed (i.i.d) random variables (r.v.) γ_i, which are not dependent on input process, model state, and have general distribution $G_i(t) = P(\gamma_i \leq t)$ with finite mean value $\overline{\gamma_i}$. The catastrophes occurrence process is a non-homogeneous MMAP $\xi_c(t)$ as well. When catastrophes occur, all customers in the model are destroyed (removed, flushed out) instantaneously and the model becomes empty. Then the model continues working from the empty state. Assume that the MMAP $\xi_0(t)$ (respectively $\xi_c(t)$) is given by an underlying Markov process (phase process (PP)) $\{J_0(t); t \geq 0\}$ $(\{J_c(t); t \geq 0\})$ with finite set of states $E_0 = \{1, 2, ..., m_0\}$ $(E_c = \{1, 2, ..., m_c\})$ and sequence of time varying characteristic matrices $\{D_{00}(t), D_{0h}(t); \mathbf{h} \in C_0^0\}$ $(\{D_{c0}(t), D_{ch}(t); \mathbf{h} \in C_c^0\})$ of $m_0 \times m_0$ $(m_c \times m_c)$ size, where C_0^0 and C_c^0 are a finite or counting sets of nonnegative integers (size of arriving batches), $\mathbf{h} = (h_1, h_2, ..., h_K)$, $\mathbf{h} \in C_0^0(C_c^0)$.

As shown in [7,9] the superposition of finite number of MMAPs is still a MMAP, so instead of using $\xi_0(t)$ and $\xi_c(t)$ MMAPs for arrival of customers and occurrence of catastrophes we will consider a superposition MMAP $\xi(t)$. The underlying PP $\{J(t); t \geq 0\}$ of superposition MMAP $\xi(t)$ is defined on the finite set of states $E = \{1, 2, ..., m\}$, $m = m_0 m_c$ and sequence of time varying characteristic matrices $\{D_0(t), D_c(t), D_\mathbf{h}(t); \mathbf{h} \in C^0\}$ of $m \times m$ size, where C^0 is a finite or counting set of nonnegative integers, $\mathbf{h} = (h_1, h_2, ..., h_K)$, $\mathbf{h} \in C^0$, and h_r is a number of type r customers, $1 \leq r \leq K$ in a batch. $D_0(t) = D_{00}(t) \oplus D_{0c}(t)$ is a non-singular matrix with negative diagonal and non-negative extra-diagonal elements, the column-wise sum of which is less than or equal to zero. $D_0(t)$ manages PP $J(t)$ transitions, that are not accompanied with customer generation or occurrence of catastrophes. Non-negative matrices $D_h(t) = D_h(t) \oplus I_{m_c}$ govern the transitions of the PP $J(t)$ with a mark $\mathbf{h} = (h_1, h_2, ..., h_K)$ which are accompanied with generation of batches of customers. A non-negative matrix $D_c(t) = I_{m_0} \oplus \sum_{\mathbf{h} \in C_c^0} D_{ch}(t)$ governs the transitions of the PP $J(t)$ which are accompanied with occurrence of catastrophes. Here $\oplus$ and $\otimes$ denote the Kronecker sum and product. Further, we assume that PP $J(t)$ is an irreducible Markov process with generating matrix $D(t)$, with a set E of states, and with distribution vector $\pi(t) = (\pi_1(t), \pi_2(t), ..., \pi_m(t))$. Here $D(t)$ is a matrix of size $m \times m$:

$$D(t) = D_0(t) + D_c(t) + \sum_{\mathbf{h} \in C^0} D_\mathbf{h}(t), (D(t) \neq D_0(t)),$$

$$\pi'(t) = \pi(t)D(t), D(t)\mathbf{e} = 0, \pi(t)\mathbf{e} = 1,$$

where $\mathbf{e}$ is a unit column vector.

The arrival rate of customers $\lambda(t)$ and occurrence rate of catastrophes $\lambda_c(t)$ are defined as

$$\lambda(t) = \pi(t) \sum_{n=1}^{\infty} n \sum_{\mathbf{h} \in C^0, h_1 + h_2 + \ldots + h_K = n} D_\mathbf{h}(t) \mathbf{e}, \quad \lambda_c(t) = \pi(t) D_c(t) \mathbf{e}.$$

.

Let us define the following random vectors:

$N(t) = (N_1(t), N_2(t), \ldots, N_K(t))$: the numbers of customers of K types of customers in the model at moment t, were $N_r(t)$ represents the number of r th type of customers in the model at moment t, $1 \leq r \leq K$.

$\mathbf{M}(t) = (M_1(t), M_2(t), \ldots, M_K(t))$: the numbers of customers of K types of customers which depart from the system in time-interval $(0, t]$ were $M_r(t)$ represents the number of r th type of customers which depart from the system in time-interval $(0, t]$, $1 \leq r \leq K$.

3 Model Analysis

Let us suppose that the model is observed during the time interval $[u, t)$. Denote by $N(u, t)$ the number of customers arrived at moment u, $0 \leq u \leq t$, and are being served at the moment t, $\mathbf{N}(t) = \mathbf{N}(0, t)$, $\mathbf{N}(0) = \mathbf{0}$, where $\mathbf{0}$ is a vector with 0 elements. Correspondingly denote by $\mathbf{M}(u, t)$ the number of customers that arrived at moment u, and have been served up to moment t, $\mathbf{M}(t) = \mathbf{M}(0, t)$, $\mathbf{M}(0) = \mathbf{0}$.

Let $R_{jk}(\mathbf{n}, u, t)$ be the probability that from the customers arrived at moment u, $0 \leq u \leq t$, $\mathbf{n} = (n_1, n_2, \ldots, n_K)$ are still in service at moment t, and PP $J(u)$ is in the phase $j \in E$ under condition that at initial moment $u = 0$ the system was empty, and PP $J(u)$ was in phase $k \in E$,

$$R_{jk}(\mathbf{n}, u, t) = P(\mathbf{N}^S(u, t) = \mathbf{n}, J(u) = j | \mathbf{N}^S(0) = \mathbf{0}, J(0) = k).$$

Let $W_{jk}(\mathbf{n}, u, t)$ be the probability that from the customers arrived at moment u, $0 \leq u \leq t$, $\mathbf{n} = (n_1, n_2, \ldots, n_K)$ have been served up to moment t, and PP $J(u)$ is in the phase $j \in E$ under condition that at initial moment $u = 0$ the system was empty, and PP $J(u)$ was in phase $k \in E$,

$$W_{jk}(\mathbf{n}, u, t) = P(\mathbf{M}(u, t), J(u) = j | \mathbf{M}(0) = \mathbf{0}, J(0) = k).$$

$R(\mathbf{n}, u, t)$, $W(\mathbf{n}, u, t)$ are matrices of size $m \times m$ with elements $R_{jk}(\mathbf{n}, u, t)$ and $W_{jk}(\mathbf{n}, u, t)$ respectively. Assume also $\mathbf{R}(u, t) = (R(\mathbf{n}, u, t), \mathbf{n}, \mathbf{n} \geq \mathbf{0})$, $\mathbf{R}(t) = \mathbf{R}(0, t)$, $\mathbf{W}(u, t) = (W(\mathbf{n}, u, t), \mathbf{n} \geq \mathbf{0})$, $\mathbf{W}(t) = \mathbf{W}(0, t)$.

Let us suppose that the model is observed at fixed moment of time t. Then a mark (batch) $\mathbf{h} = (h_1, h_2, \ldots, h_K)$ of customers arrives during infinitesimal time interval du, $(u, u + du)$, $u \in [0, t)$ with the rate of $D_{\mathbf{h}(u)}$ according to the MMAP arrival process $\xi(t)$. It is well known that for given mark $\mathbf{h} = (h_1, h_2, \ldots, h_K)$ the probability of $n_r = (0, 1, \ldots, h_r)$ customers of type r still being served (have been served) at time t is distributed binomially by

$$b_{h_r}(n_r, t - u) = \binom{h_r}{n_r}(1 - G_r(t - u))^{n_r} G_r(t - u)^{h_r - n_r}, 0 \le n_r \le h_r. \quad (1)$$

$$(d_{h_r}(n_r, t - u) = \binom{h_r}{n_r} G_r(t - u)^{n_r}(1 - G_r(t - u))^{h_r - n_r}, 0 \le n_r \le h_r.) \quad (2)$$

Conditioning upon the size of the marks arrival, the total rate of $\mathbf{n} = (n_1, n_2, ..., n_K)$ customer arrival during the time interval du which are still in service (have been served) at time t is

$$\begin{aligned}
K_{\mathbf{n}}(u, t) &= \sum_{\mathbf{h}=\mathbf{n}}^{\infty} D_{\mathbf{h}}(u) \prod_{r=1}^{K} b_{h_r}(n_r, t - u) \\
&= \sum_{\mathbf{h}=\mathbf{n}}^{\infty} D_{\mathbf{h}}(u) \prod_{r=1}^{K} \binom{h_r}{n_r}(\overline{G}_r(t - u))^{n_r} G_r(t - u)^{h_r - n_r}.
\end{aligned} \quad (3)$$

Where $\overline{G}(t) = 1 - G(t)$.

$$\left(\begin{aligned}
V_{\mathbf{n}}(u, t) &= \sum_{\mathbf{h}=\mathbf{n}}^{\infty} D_{\mathbf{h}}(u) \prod_{r=1}^{K} d_{h_r}(n_r, t - u) \\
&= \sum_{\mathbf{h}=\mathbf{n}}^{\infty} D_{\mathbf{h}}(u) \prod_{r=1}^{K} \binom{h_r}{n_r} G_r(t - u)^{n_r}(\overline{G}_r(t - u))^{h_r - n_r}
\end{aligned}\right)$$

The probabilities $R(\mathbf{n}, u, t), (W(\mathbf{n}, u, t))$ satisfy the following systems of differential equations

$$\begin{aligned}
\frac{\partial}{\partial u} R(\mathbf{0}, u, t) &= [D_0(u) + K_0(u, t)] R(\mathbf{0}, u, t) + D_c(u) \sum_{k=0}^{\infty} R(\mathbf{k}, u, t), \\
\frac{\partial}{\partial u} R(\mathbf{n}, u, t) &= [D_0(u) + K_0(u, t)] R(\mathbf{n}, u, t) + \sum_{k \neq 0}^{n} R(\mathbf{k}, u, t) K_{n-k}(u, t),
\end{aligned} \quad (4)$$

$$\begin{aligned}
\frac{\partial}{\partial u} W(\mathbf{0}, u, t) &= [D_0(u) + V_0(u, t)] W(\mathbf{0}, u, t), \\
\frac{\partial}{\partial u} W(\mathbf{n}, u, t) &= [D_0(u) + V_0(u, t)] W(\mathbf{n}, u, t) + \sum_{k \neq 0}^{n} W(\mathbf{k}, u, t) K_{n-k}(u, t),
\end{aligned} \quad (5)$$

with initial conditions $R(0, 0, t) = 1$, $R(\mathbf{n}, u, t) = \mathbf{0}$, $\mathbf{W}(\mathbf{0}, 0, t) = \mathbf{1}$, $\mathbf{W}(\mathbf{n}, u, t) = \mathbf{0}$.

Let us denote by $\tilde{A}(z, u, t)$ the generating functions of $A(\mathbf{n}, u, t)$

$$\tilde{A}(z, u, t) = \sum_{n=0}^{\infty} z^n A(\mathbf{n}, u, t), |z_1| \le 1, |z_2| \le 1, ..., |z_K| \le 1,$$

where $\mathbf{z^n} = z_1^{n_1} z_2^{n_2} ... z_K^{n_K}$.

Then for $R(\mathbf{n}, u, t)$, and $W(\mathbf{n}, u, t)$, we define the corresponding PGFs

$$\tilde{R}(z, u, t) = \sum_{n=0}^{\infty} z^n R(\mathbf{n}, u, t), \tilde{W}(z, u, t) = \sum_{n=0}^{\infty} z^n W(\mathbf{n}, u, t), |z_1| \le 1, ..., |z_K| \le 1.$$
(6)

The PGFs $\tilde{R}(z, u, t)$ and $\tilde{W}(z, u, t)$ satisfy the following differential equations

$$\frac{\partial}{\partial u} \tilde{R}(z, u, t) = [D_0(u) + \tilde{K}(z, u, t)]\tilde{R}(z, u, t) + D_c(u)\tilde{R}(1, u, t),$$
(7)

$$\frac{\partial}{\partial u} \tilde{W}(z, u, t) = [D_0(u) + \tilde{V}(z, u, t)]\tilde{W}(z, u, t),$$
(8)

with initial conditions $\tilde{R}(z, 0, t) = \mathbf{I}$, $\tilde{W}(z, u, t) = \mathbf{I}$.
Here

$$\tilde{K}(z, u, t) = \sum_{n=0}^{\infty} D_n(u) \prod_{r=1}^{K} [z_r \overline{G}_r(t - u) + G_r(t - u)]^{n_r},$$

$$\tilde{V}(z, u, t) = \sum_{n=0}^{\infty} D_n(u) \prod_{r=1}^{K} [z_r G_r(t - u) + \overline{G}_r(t - u)]^{n_r},$$

The solutions of Eqs. (7) and (8) can be presented in matrix-exponential form

$$\tilde{R}(z, u, t) = e^{\int_0^u [D_0(x) = \tilde{K}(z, x, t)]dx} + \int_0^u e^{\int_x^u [D_0(x) + \tilde{K}(z, x, t)dx]} D_c(x)\tilde{R}(1, x, t)dx.$$
(9)

$$\tilde{W}(z, u, t) = e^{\int_0^u [D_0(x) + \tilde{V}(z, x, t)]dx}.$$

Where $\tilde{R}(1, u, t)$ satisfy the differential equation

$$\frac{\partial}{\partial u} \tilde{R}(1, u, t) = [D_0(u) + \tilde{K}(1, u, t)]\tilde{R}(1, u, t),$$
(10)

with initial condition $\tilde{R}(1, 0, 0) = \mathbf{I}$.

$$\tilde{R}(1, u, t) = e^{\int_0^u D(x)dx}$$
(11)

From (7) it follows that for homogeneous model when $D_n(u) = D_n, u < t$, the rates $\tilde{K}(z, u, t), \tilde{V}(z, u, t)$ are independent of $t, \tilde{K}(z, u) = \tilde{K}(z, t - u, t)$, $\tilde{V}(z, t) = \tilde{V}(z, t - u, t)$. Hence the solutions of Eqs. (7) and (8) can be presented in matrix-exponential form

$$\tilde{R}(z, t) = e^{\int_0^t [D_0 + \tilde{K}(z, x)]dx} + \int_x^t e^{\int_x^t [D_0 + \tilde{K}(z, x)]dx} D_c(x)\tilde{R}(1, x)dx, \quad |z| \le 1 \quad (12)$$

$$\tilde{W}(z, t) = e^{\int_0^t [D_0 + \tilde{V}(z, x)]dx}, \quad |z| \le 1$$

If we suppose that at time $t = 0$, there are $\mathbf{h_0} = (h_{01}, h_{02}, ..., h_{0K})$ initial customers in the model then for $\tilde{R}(z,t)$ and $\tilde{W}_{h_0}(z,t)$ we get

$$\tilde{R}_{h_0}(z,t) = \prod_{r=1}^{K} [z_r \overline{G}_r(t) + G_r(t)]^{h_r} e^{\int_0^t [D_0 + \tilde{K}(z,u)]du}, \quad |z| \leq 1$$

$$\tilde{W}_{h_0}(z,t) = \prod_{r=1}^{K} [z_r \overline{G}_r(t) + G_r(t)]^{h_r} e^{\int_0^t [D_0 + \tilde{V}(z,u)]du}, \quad |z| \leq 1.$$

(13)

Remark 1. Let us consider the service policy when we do not distinguish between the customers in the same mark, and they are all served together as one customer with the service time $\gamma = max\{\gamma_1, \gamma_2, ..., \gamma_K\}$ with the distribution function $G(t)$ and mean value $\overline{\gamma}_1$.

$$G(t) = \prod_{i=1}^{K} G_i(t).$$

In this case the MMAP transforms to ordinary MAP (see for example in [27,28]) with coefficient matrices $\{D_0, D_c, \tilde{D}_1\}$, where $\tilde{D}_1 = \sum_{\mathbf{h} \in C^0} D_{\mathbf{h}}$. Denote by $N(t)$ the number of arrivals of batches in $(0, t]$ and $J(t)$ the PP of underlying Markov chain with generator $D = D_0 + D_c + \tilde{D}_1$ matrix. Denote by $R(n,t)$, $n \geq 0$ an $m \times m$ matrices with (j,k) th element $R_{j,k}(n,t) = P\{N^S(t) = n, J(t) = j | N^s(0) = 0, J(0) = k\}$ as the conditional probability of having n batches in service at time t and PP is in state j when at initial moment t the model was empty and PP was in state k. These probabilities satisfy the system of differential equations

$$\frac{d}{dt} R(0,t) = R(0,t)[D_0 + \tilde{D}_1 G(t)] + D_c \sum_{k=0}^{\infty} R(k,t),$$

$$\frac{d}{dt} R(n,t) = R(n,t)[D_0 + \tilde{D}_1 G(t)] + R(n-1,t)\tilde{D}_1(t - G(t)), n \geq 1.$$

With initial conditions $R(0,0) = \mathbf{1}, R(n,0) = \mathbf{0}$.

Let us denote the PGF of $\{R(n,t), n \geq 0\}$ by $\tilde{R}(z,t)$. Then $\tilde{R}(z,t)$ satisfies the following differential equation

$$\frac{d}{dt} \tilde{R}(z,t) = \tilde{R}(z,t)[D_0 + \tilde{D}_1(G(t) + z(t - G(t)))] + D_c \tilde{R}(1,t), \quad |z| \leq 1,$$

with initial conditions $\tilde{R}(1,0) = I$.

$$\tilde{R}(z,t) = e^{\int_0^t [D_0 + \tilde{D}_1(G(x) + z(1 - G(x)))]dx}$$

$$+ \int_0^t e^{\int_0^t [D_0 + \tilde{D}_1(G(x) + z(1 - G(x)))]dx} D_c \tilde{R}(1,u)dx.$$

(14)

Remark 2. Now we consider the model $MMAP|G|\infty$ when catastrophes occur according to transient Poisson distribution with parameter $v(t)$. In this case the probabilities $R(\mathbf{n,u},t)$ satisfy the differential equations

$$\frac{d}{du}R(\mathbf{0},u,t) = [D_0(u) + K_0(u,t) - v(u)I]R(\mathbf{0},u,t) + v(u)\sum_{k=0}^{\infty} R(\mathbf{k},u,t),$$

$$\frac{d}{du}R(\mathbf{n},u,t) = [D_0(u) + K_0(u,t) - v(u)I]R(\mathbf{n},u,t) + \sum_{\mathbf{n}\neq\mathbf{k}} R(\mathbf{k},u,t)K_{\mathbf{n-k}}(u,t)$$

$$(15)$$

with initial conditions $R(\mathbf{0},u,0) = \mathbf{1}$ and $R(\mathbf{n},u,0) = \mathbf{0}$, where I, $\mathbf{1}$ and $\mathbf{0}$ are identity, unite and null matrices.

Then from (15) for the PGF $\tilde{R}(z,u,t)$ of the number of customers being in service at moment t we derive

$$\frac{d}{du}\tilde{R}(z,u,t) = (D_0(u) + \tilde{K}(z,u,t) - v(u)I)\tilde{R}(z,u,t) + v(u)\tilde{R}(\mathbf{1},u,t) \quad (16)$$

with initial conditions $\tilde{R}(z,0,t) = I$.

Hence the solution of Eq. (16) can be presented in matrix-exponential form

$$\tilde{R}(z,t) = e^{\int_0^t [D_0(u) + \tilde{K}(z,u,t) - v(u)I]du}$$

$$+ \int_0^t v(u)e^{\int_u^t [D_0(u) + \tilde{K}(z,u,t) - v(u)I]du}\tilde{R}(\mathbf{1},u)du, |z| \leq 1, \quad (17)$$

where the matrix $\tilde{R}(\mathbf{1},t)$ is defined by the differential equation

$$\frac{d}{du}\tilde{R}(\mathbf{1},u,t) = D(u)\tilde{R}(\mathbf{1},u,t) \quad (18)$$

and has a solution

$$\tilde{R}(\mathbf{1},u,t) = e^{\int_u^t D(x)dx}.$$

Remark 3. Let us consider a homogeneous model $MAP|G|\infty$ with MAP arrival of customers and non-stationary Poisson occurrence of catastrophes with rate $v(t)$. Suppose that MAP is given by characteristic matrices C, D size of $m \times m$. Then for the PGF of this model $\tilde{R}(z,t)$ we have

$$\tilde{R}(z,t) = e^{\int_0^t \{C+D[G(x)=z(1-G(x))]-v(x)I\}dx}$$

$$\left(I + \int_0^t v(y)e^{\int_0^y \{D(1-z)(1-G(x))+v(x)Idx\}}dy\right). \quad (19)$$

Thus for probabilities $R(n,t)$ we obtain

$$R(u,t) = e^{\int_0^t \{C+DG(x)-v(x)I\}dx}\frac{\left(\int_0^t D\overline{G}(x)dx\right)^n}{x!} +$$

$$\int_0^t v(y)e^{[(C+D)t-\int_y^t(D(1-G(x))+v(x)I)dx]}\frac{\left(\int_y^t D\overline{G}dx\right)^n}{n!}dy.$$

For the first and second moments $m_1(t), m_2(t)$ of the number of busy servers we obtain

$$\frac{d}{dt}m_1(t) = [C + D - v(t)]m_1(t) + D(1 - G(t))e^{(C+D)t},$$

$$\frac{d}{dt}m_2(t) = [C + D - v(t)]m_2(t) + 2D(1 - G(t))m_1(t).$$

with the initial conditions $m_1(0), m_2(0) = 0$.

$$m_1(t) = \int_0^t e^{\int_y^t [C+D-v(x)I]dx} De^{[C+D]y}(1 - G(y))dy,$$

$$m_2(t) = 2\int_0^t e^{\int_y^t [C+D-v(x)I]dx} Dm_1(t)(1 - G(y))dy. \tag{20}$$

Remark 4. Now let us the model $MMAP_k(t)|G_k|\infty$ with MAP arrival of catastrophes. Suppose that the arrivals of customers and occurrence of catastrophes are given by corresponding matrices $\{D_0(t), D_h(t)\}$ and $\{C_2(t), D_2(t)\}$ respectively. Then from (15) for the PGF of numbers of customers in the model $\tilde{R}(z, u, t)$ at moment t satisfies the following differential equation

$$\frac{d}{du}\tilde{R}(z, u, t) = [D_0(u) + C_2(u) + \tilde{K}(z, u, t)]\tilde{R}(z, u, t) + D_2(u)\tilde{R}(1, u, t) \tag{21}$$

with initial conditions $\tilde{R}(z, 0, t) = I, R(\tilde{z}, 0, t) = I$.

The solution of Eq. (21) can be presented in matrix-exponential form

$$\tilde{R}(z, u, t) = e^{\int_0^u [D_0(x)+\tilde{K}(z,x,t)+C_2(x)]dx}$$

$$+ \int_0^u e^{\int_y^u [D_0(x)+\tilde{K}(z,x,t)+C_2(x)]dx} D_2(y)\tilde{R}(1, y, t)dy. \tag{22}$$

From (22), the PGFs of the model $BMAP_k|G_k|\infty$ without catastrophes (see in [13]) and the model $M|M|\infty$ with catastrophes (see in [15]) can be derived as particular cases.

$$\tilde{R}(z, t) = e^{\int_0^t [D_0+\tilde{K}(z,x)+C_2(x)]dx} + \int_0^t e^{\int_y^t [D_0+\tilde{K}(z,x)+C_2(x)]dx} D_2(y)\tilde{R}(1, y)dy. \tag{23}$$

Remark 5. In this part we apply the model of a queue with random volume of customers [35] to the infinite-server queue $MMAP_k|G_k|\infty$ queue. Afterward we will use the definitions and notations of [35]. Let's suppose that demands of r type customers are characterized with random volume ζ_r and service time ξ_r, which are i.i.d. and mutually independent r.v.s. Let us also denote:

$F_r(x, t) = P(\zeta_r < z, \xi_r < t)$ is the joint distribution function of the demand volume and service time;

$L_r(x) = F_r(x, \infty)$ is the demand volume distribution function;

$G_r(t) = F_r(\infty, t)$ is the demand service time distribution function;

$\sigma(t)$ is the total volume of the demand being in service at moment t;

$D(x,t)$ is the distribution function of $\sigma(t)$;

$\delta(s,t)$ is the be the LST of $D(x,t)$ for the model without catastrophes;

$\delta_v(s,t)$ is the be the LST of $D(x,t)$ for the model with catastrophes;

$$\delta(s,t) = \int_0^\infty e^{-sx}dD(x,t),$$

$$\delta(s,q) = \int_0^\infty \int_0^\infty e^{-(sx+qt)}dD(x,t),$$

$$\alpha(s_r,q_r,t) = \int_0^\infty \int_0^t e^{-(s_r x + q_r u)}dF_r(x,u),$$

$\alpha(s_r,q_r,\infty)$ is the LST of $F_r(x,u)$.

Then for LST of total volume distribution $\delta(s,q)$ of the model $MMAP|G|\infty$ by means of results in [40] Theorem 1 we derive

$$\tilde{\delta}(s,t) = exp\{\int_0^t \tilde{\epsilon}(s,u,t)du\},$$

$$\tilde{\epsilon}(s,u,t) = \sum_{n=0}^\infty D_n \prod_{r=1}^K \gamma^{n_r}(s_r,u,t), \tag{24}$$

$$\gamma(s_r,u,t) = G_r(u) + \overline{G}_r(u)\chi(s_r,t) = G_r(u) + \frac{\overline{G}_r(u)}{\beta_{r1}(t)}\frac{\partial}{\partial q_r}\alpha(s_r,q_r,t)\Big|_{q_r=0},$$

$$\chi(s_r,t) = \frac{1}{\beta_{1r}(t)}\int_0^\infty \int_0^t e^{-(s_r x + q_r u)}u\,dF_r(x,u) = \frac{1}{\beta_{1r}}\frac{\partial}{\partial q_r}\alpha(s_r,q_r,t)\Big|_{q_r=0},$$

$$\beta_{1r}(t) = \int_0^t u\,dF_r(\infty,u) = \int_0^t \beta_r(u).$$

For the model $BMAP|G|\infty$ with the Poisson stream of catastrophes with the rate $v(t)$ and characteristic arrival matrices $\{D_n, n \geq 0\}$ of customers the LST and LT of the total volume at the moment t is given by

$$\tilde{\epsilon}(s,u,t) = D_0 + \sum_{n=1}^\infty D_n \left[G_r(u) + \frac{\overline{G}_r(u)}{\beta_{r1}(t)}\frac{\partial}{\partial q_r}\alpha(s_r,q_r,t)\Big|_{q_r=0} \right]^n$$

$$= D_0 + \hat{D}\left[G(u) + \frac{\overline{G}(u)}{\beta_1(t)}\frac{\partial}{\partial q}\alpha(s,q,t)\Big|_{q=0} \right],$$

$$\tilde{\tilde{\delta}}(s,q) = \int_0^\infty e^{-qIt + \int_0^t \{D_0 + \hat{D}[G(u) + \frac{1-G(u)}{\beta_1(t)}\frac{\partial}{\partial q}\alpha(s,q,t)|_{q=0}]\}du}dt, \tag{25}$$

$$\delta_v(s,t) = e^{\int_0^t \{D_0 + \hat{D}[G(u) + \frac{\overline{G}(u)}{\beta_1(t)}\frac{\partial}{\partial q}\alpha(s,q,t)|_{q=0}] - v(u)I\}du}$$

$$+ \int_0^t v(u)e^{\int_0^t \{D_0 + \hat{D}[G(x) + \frac{\overline{G}(x)}{\beta_1(t)}\frac{\partial}{\partial q}\alpha(s,q,t)|_{q=0}] - v(x)I\}dx}du.$$

In case of $v(t) = v$ for LST and LT of the total volume and its mean value of the model with catastrophes we derive

$$\tilde{\tilde{\delta}}_v(s,q) = \left(1 + \frac{v}{q}\right)\tilde{\tilde{\delta}}(s, q+v),$$

$$\tilde{\delta}_v(s) = v\tilde{\tilde{\delta}}(s,v).$$

(26)

4 Conclusion

In this paper, we considered the time inhomogeneous infinite-server queue model $MMAP_k(t)|G_k|\infty$ with marked MAP arrival of customers and catastrophes. We suppose that each customer is characterized by the random amount of service time and random volume. By using MMAP thinning method, the transient joint PGFs of queue length of different types of customers present in the model at moment t and the number of different types of customers departing from the system in time interval $(0, t]$ are found. The basic differential equations for corresponding PGFs for transient and stationary models are obtained. The LST of total volume of customers and its mean value for the models with time varying MMAP and Poisson stream of catastrophes are obtained.

The obtained results may be applied for performance evaluation, as well as designing the optimal strategies for managing resources of a wide class of systems and networks, whereas the model $MMAP_k(t)|G_k|\infty$ may be used as a mathematical model of their subsystems.

Acknowledgement. This work was supported by "Data Science Program with Career Support and Connections to Industry," NSF Award 1842386 grant.

References

1. Paxson, V., Floyd, S.: Wide-area traffic: the failure of Poisson modeling. In: Proceedings of the ACM, pp. 257–268 (1994)
2. Neuts, M.F.: A versatile Markovian point process. J. App. Prob. **16**(4), 764–779 (1979)
3. He, Q.-M.: Queues with marked customers. Adv. App Prob. **30**, 365–372 (1996)
4. Cordeiro, J.D., Kharoufeh, J.P.: Batch Markovian Arrival Processes (BMAP). In: Cochran, J., Cox, T., Keskinocak, P., Kharoufeh, J.P., Smith, J.C. (eds.) Wiley Encyclopedia of Operations Research and Management Science. Wiley, New York (2011)
5. Chakravarthy, S.R.: The batch Markovian arrival process: a review and future work. In: Krishnamoorthy, A., Raju, N., Ramaswami, V. (eds.) Advances in Probability Theory and Stochastic Processes, pp. 21–39. Notable Publications, New York City (2000)
6. Artalejo, J.R., Gomez-Corral, A., He, Q.M.: Markovian arrivals in stochastic modelling: a survey and some new results. SORT **34**(2), 101–144 (2010)
7. Pacheco, A., Tang, C.H.L., Prabhu, N.U.: Markov-Additive Processes and Semi-regenerative Phenomena. World Scientific, Singapore (2009)

8. Breuer, L.: Introduction to stochastic processes. University of Kent (2014)
9. He, Q.: Fundamentals of Matrix-Analytic Methods. Springer, New York (2014). https://doi.org/10.1007/978-1-4614-7330-5
10. Shanbhag, D.N.: On innite-server queues with batch arrivals. J. Appl. Prob. **9**, 208–213 (1966)
11. Brown, M., Ross, S.: Some results for infinite server poisson queues. J. Appl. Prob **6**, 604–611 (1969)
12. Postan, M.Ya.: Flow of serviced requests in infinite-channel queueing systems in a transient mode. Probl. Inform. Trans. **13**(4), 309–313 (1977)
13. Eick, S.G., Massey, W.A., Whitt, W.: The physics of the $M(t)|G|\infty$ queue. Oper. Res. **41**, 731–742 (1993)
14. Massey, W.A.: The analysis of queues with time-varying rates for telecommunication models. Telecom. Syst. **21**(2–4), 173–204 (2002)
15. Whitt, W.: Heavy-traffic fluid limits for periodic infinite-server queues. Queueing Syst. **84**, 111–143 (2016)
16. Economou, A., Fakinos, D.: Alternative approaches for the transient analysis of Markov chains with catastrophes. J. Stat. Theory Pract. **2**(2), 183–197 (2008). https://doi.org/10.1080/15598608.2008.10411870
17. Kerobyan, K.: Infinite-server $M|G|\infty$ queueing models with catastrophes, 12 December 2018. http://www.arxiv.org/ftp/arxiv/papers/1807/1807.08609.pdf
18. Kerobyan, K., Kerobyan, R.: Transient analysis of infinite-server queue $MMAP_k(t)|G_k|\infty$ with marked MAP arrival and disasters. In: Proceedings of 7th International Conference on "HET-NETs 2013", November 2013, Ilkley, UK, pp. 11–13 (2013)
19. Kerobyan, R., Kerobyan, K., Covington, R.: Infinite-server queue model $MMAP_k(t)|G_k|\infty$ with time varying marked MAP arrivals and catastrophes. In: Proceedings of the 18th International Conference on Named Aafter A. F. Terpugov, Information Technologies ad Mathematical Modelling, ITMM , 26–30 June 2019, Saratov, Russia (2019)
20. Pang, G., Whitt, W.: Innite-server queue with batch arrivals and dependent service times. Prob. Eng. Inf. Sci. **26**, 197–220 (2012)
21. Ramaswami, V., Neuts, M.F.: Some explicit formulas and computational methods for infinite-server queues with phase-type arrival. J. Appl. Prob. **17**, 498–514 (1980)
22. Ramaswami, V.: The $N/G/\infty$ queue. Technical report, Department of Mathematics, Drexel University, Philadelphia, PA (1978)
23. Latouche, G., Ramaswami, V.: Introduction to Matrix Analytic Methods in Stochastic Modeling. SIAM, Philadelphia (1999)
24. Moiseev, A., Nazarov, A.: Infinite-server Queueing Systems and Networks. Publ. NTL, Tomsk (2015)
25. Lisovskaya, E., Moiseeva, S., Pagano, M.: Multiclass GI/GI/∞ queueing systems with random resource requirements. In: Dudin, A., Nazarov, A., Moiseev, A. (eds.) ITMM/WRQ -2018. CCIS, vol. 912, pp. 129–142. Springer, Cham (2018). https://doi.org/10.1007/978-3-319-97595-5_11
26. Masuyama, H.: Studies on algorithmic analysis of queues with batch Markovian arrival streams. Ph.D., Thesis, Kyoto University (2003)
27. Nazarov, A., Moiseeva, S.: Asymptotic Analysis Method in Queueing Theory. NTL, Tomsk (2006)
28. Moiseev, A.: Asymptotic analysis of queueing system $MAP/GI/\infty$ with high-rate arrivals. Tomsk State Univ. J. Control Comput. Sci. **3**(32), 56–65 (2015)
29. D'Auria, B.: $M/M/\infty$ queues in semi-Markovian random environment. Queueing Syst. **58**, 221–237 (2008)

30. Fralix, B.H., Adan, I.J.B.F.: An infinite-server queue influenced by a semi-Markovian environment. Queueing Syst. **61**, 65–84 (2009)
31. Linton, D., Purdue, P.: An $M|G|\infty$ queue with m customer types subject to periodic clearing. Opsearch **16**, 80–88 (1979)
32. Purdue, P., Linton, D.: An infinite-server queue subject to an extraneous phase process and related models. J. Appl. Prob. **18**, 236–244 (1981)
33. Kerobyan, K., Covington, R., Kerobyan, R., Enakoutsa, K.: An infinite-server queueing $MMAP_k|G_k|\infty$ model in semi-markov random environment subject to catastrophes. In: Dudin, A., Nazarov, A., Moiseev, A. (eds.) ITMM/WRQ -2018. CCIS, vol. 912, pp. 195–212. Springer, Cham (2018). https://doi.org/10.1007/978-3-319-97595-5_16
34. Kerobyan, K., Kerobyan, R., Enakoutsa, K.: Analysis of an infinite-server queue $MAP_k|G_k|\infty$ in random environment with k markov arrival streams and random volume of customers. In: Dudin, A., Nazarov, A., Moiseev, A. (eds.) ITMM/WRQ -2018. CCIS, vol. 912, pp. 305–320. Springer, Cham (2018). https://doi.org/10.1007/978-3-319-97595-5_24
35. Tikhonenko, O.M.: Queueing models in computer systems. Universitetoe, Minsk (1990)
36. Tikhonenko, O.M.: Computer Systems Probability Analysis. Akademicka Oficyna Wydawnicza EXIT, Warsaw (2006)
37. Tikhonenko, O.M., Tikhonenko-Kędziak, A.: Multi-server closed queueing system with limited buffer size. J. Appl. Math. Comp. Mech. **16**(1), 117–125 (2017)
38. Tikhonenko, O.M.: Basics of queueing theory. Lecture notes, TSU (2013)
39. Moiseev, A., Moiseeva, S., Lisovskaya, E.: Infinite-server queueing tandem with MMPP arrival and random capacity of customers. In: Proceedings of the 31st European Conference on Modelling and Simulation Budapest, Hungary 23–26 May 2017
40. Naumov, V., Samuylov, K.: On the modeling of queue systems with multiple resources. Proc. RUDN. **3**, 60–63 (2014)

Performance Evaluations of a Cloud Computing Physical Machine with Task Reneging and Task Resubmission (Feedback)

Godlove Suila Kuaban[1], Bhavneet Singh Soodan[2](✉), Rakesh Kumar[2](✉), and Piotr Czekalski[3](✉)

[1] Institute of Theoretical and Applied Informatics, Polish Academy of Sciences, Baltycka 5, 44-100 Gliwice, Poland
gskuaban@iitis.pl
[2] School of Mathematics, Shri Mata Vaishno Devi University, Katra 182320, Jammu and Kashmir, India
bhavneet5678@gmail.com, rakesh.kumar@smvdu.ac.in
[3] Department of Computer Graphics, Vision and Digital Systems, Faculty of Automatic Control, Electronics and Computer Science, Silesian University of Technology, Akademicka 16, 44-100 Gliwice, Poland
piotr.czekalski@polsl.pl

Abstract. Cloud service providers (CSP) provide on-demand cloud computing services, reduces the cost of setting-up and scaling-up IT infrastructure and services, and stimulates shorter establishment times for start-ups that offer or use cloud-based services. Task reneging or dropping sometimes occur when a task waits in the queue longer than its timeout or execution deadline, or it is compromised and must be dropped from the queue or as an active queue management strategy to avoid tail dropping of tasks when the queues are full. Reneged or dropped tasks could be resubmitted provided they were not dropped due to security reasons. In this paper, we present a simple M/M/c/N queueing model of a cloud computing physical machine, where the interarrival times and the services times are exponentially distributed, with N buffer size and c virtual machines running in parallel. We present numerical examples to illustrate the effect of task reneging and task resubmission on the queueing delay, probability of task rejection, and the probability of immediate service.

Keywords: Transient-state · Steady-state · Performance evaluations · Cloud computing · Physical machines · Tasks reneging or dropping · Tasks resubmission or feedback

1 Introduction

Cloud service providers (CSP) provide on-demand cloud computing services such as software, platform and infrastructure to their customers. It enables the users

© Springer Nature Switzerland AG 2020
P. Gaj et al. (Eds.): CN 2020, CCIS 1231, pp. 185–198, 2020.
https://doi.org/10.1007/978-3-030-50719-0_14

to access these services anywhere, at any time and based on their needs without being concerned about the cost and time of setting up and running their infrastructure from scratch. Therefore, cloud computing has stimulated shorter establishment times for start-ups that offer or use cloud-based services and the creation of scalable enterprise applications [1]. Performance evaluations of cloud computing systems have been studied using queueing theory in [6,9–13]. The use of analytical modelling methods offer faster and less expensive performance evaluation tools when compared to testbed experiments and discrete event simulation, which are time-consuming and expensive [14]. The results obtained using analytical modelling may be an approximation of the relative trends of the performance parameters but can be used to derive high-level insight into the behaviour of the system [2]. The evaluation of cloud computing systems may require the prediction and estimation of the cost-benefit of a strategy and the corresponding acceptable quality of service (QoS) which may not be feasible by simulation or measurements [3].

Task reneging or dropping sometimes occur when a task waits in the queue longer than its timeout or execution deadline, or it is compromised and must be dropped from the queue or as an active queue management strategy to avoid tail dropping of tasks when the queues are full. Reneged or dropped tasks could be resubmitted provided they were not dropped due to security reasons. Dropping of tasks from the queue is called task reneging [15] while the resubmission of the dropped task is called feedback [16]. The authors in [4,5,17,18] studied task reneging in the context of cloud computing but their studies were limited to steady-state Markovian modelling without resubmission.

In this paper, we present a simple M/M/c/N queueing model of a cloud processing physical machine, where the interarrival times and the services times are exponentially distributed, with N buffer size and c virtual machines running in parallel. We present numerical examples to illustrate the effect of reneging and feedback on the queueing delay, probability of task rejection, and the probability of immediate service. The rest of the paper is arranged as follows: model description is presented in Sect. 2, performance modelling is presented in Sect. 3, some numerical examples are presented in Sect. 4 and conclusion in Sect. 5.

2   Model Description

The tasks submitted to a cloud computing infrastructure may be queued up in the load balancer and then scheduled to any of the available physical machines, provided the rate of arrival of tasks is far greater than the scheduling rate [7]. The load balancing mechanism detects the physical machines that are overloaded and those that are underutilised and strive to balance the load among them [7]. However, the evaluation of the load balancer is out of the scope of this paper. In the physical machines, the tasks can then be scheduled into any available VMs for processing. Because some of the tasks may be time-constrained or likely to fail or maybe have been compromised, it will renege or dropped from the queue or moved to another queue (jockeying) [8] depending on the queue management strategy implemented. Figure 1 shows a general cloud computing model where users can submit tasks over the internet to a cloud computing data centre infrastructure, which consists of the load balancer and physical machines which are hosting virtual machines.

The use of effective tasks scheduling policies ensures that the potential of cloud computing is fully harnessed and exploited to meet the QoS requirements of cloud computing services. The authors in [20] presented a review of cloud computing scheduling methods which are categorised into QoS-based task scheduling, ant Colony optimisation Algorithm-based scheduling, particle swarm optimisation (PSO)-based task scheduling, Multiprocessor-based scheduling, Fuzzy-based scheduling, Clustering-based, task scheduling, Deadline-constrained scheduling, Cost-based, scheduling and other scheduling-based approaches. Scheduling algorithms which use techniques such as round-robin, allocation, a probabilistic allocation that seek to minimise the average response time, Random Neural Network (RNN) based allocation scheme that uses reinforcement learning and on-line greedy adaptive algorithm were presented in [22]. A discrete symbiotic organism search (DSOS) scheduling algorithm was proposed in [21] for optimal scheduling of tasks in cloud data centres.

Suppose that the tasks scheduled to a given physical machine are arriving with an arrival rate of λ as shown in Fig. 2. If the rate of arrival of tasks is greater than the rate at which they are processed, then those that arrive and when all the virtual machines (VMs), $\{VM_1, VM_2, VM_3, \cdots, VM_c\}$ that are running in parallel are busy, will have to wait and then later scheduled for execution. The processing server or physical machine is modelled as an M/M/c/N, where c is the number of VMs and N is the maximum number of tasks or the buffer size. It is assumed that all the VMs have the same processing rate, μ.

The model proposed in the paper are based on the following assumptions:

1. The arrival process of tasks into the task buffers in the processing servers follows a Poisson process with parameter λ.
2. The system has a single queue and finitely many numbers of VMs. The processing times of each VM are exponentially distributed with parameter μ.
   The mean processing rate of tasks is: $\mu_n = \begin{cases} n\mu, & 0 \leq n < c \\ c\mu & c \leq n \leq N \end{cases}$
3. The queue discipline is FCFS.
4. The capacity of the system is finite (say, N).
5. The reneging times or the times at which the tasks are dropped from the queue are exponentially distributed with parameter ξ.
6. When a task reneges or is dropped from the queue, it can be resubmitted with a probability p otherwise, with a probability $q = 1 - p$ it is not resubmitted.

3   Performance Evaluation Modelling: Steady-State and Transient-State Solutions

Defining the following probabilities:

$P_0(t)$ is the probability that at time t there is no task in the system.

$P_n(t)$ is the probability that at time t there are $1 \leq n \leq N$ tasks in the system.

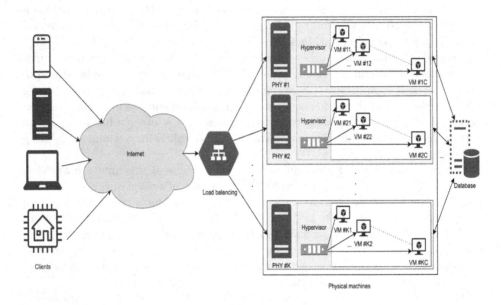

Fig. 1. Cloud computing model

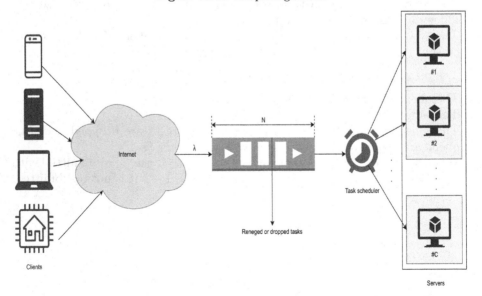

Fig. 2. Queueing model of a cloud computing physical machine

The difference-differential equations of the queuing model are:

$$\frac{dP_0(t)}{dt} = -\lambda P_0(t) + q\mu P_1(t), n = 0 \tag{1}$$

$$\frac{dP_n(t)}{dt} = -(\lambda + nq\mu)P_n(t) + \lambda P_{n-1}(t) + (n+1)q\mu P_{n+1}(t), 1 \leq n < c \tag{2}$$

$$\frac{dP_n(t)}{dt} = -(\lambda + cq\mu)P_n(t) + \lambda P_{n-1}(t) + (cq\mu + \xi)P_{n+1}(t), n = c \tag{3}$$

$$\frac{dP_n(t)}{dt} = -[\lambda + cq\mu + (n-c)\xi]P_n(t) + \lambda P_{n-1}(t)$$
$$+ [cq\mu + (n+1-c)\xi]P_{n+1}(t), c+1 \leq n < N \tag{4}$$

$$\frac{dP_N(t)}{dt} = -[cq\mu + (N-c)\xi]P_N(t) + \lambda P_{N-1}(t), n = N \tag{5}$$

In steady-state, when $\lim_{t\to\infty} P_0(t) = P_0$, $\lim_{t\to\infty} p_n(t) = p_n$, $\lim_{t\to\infty} p_N(t) = p_N$, Eqs. (1)–(5) becomes:

$$0 = -\lambda P_0 + q\mu P_1, n = 0 \tag{6}$$
$$0 = -(\lambda + nq\mu)P_n + \lambda P_{n-1} + (n+1)q\mu P_{n+1}, 1 \leq n < c \tag{7}$$
$$0 = -(\lambda + cq\mu)P_n + \lambda P_{n-1} + (cq\mu + \xi)P_{n+1}, n = c \tag{8}$$
$$0 = -[\lambda + cq\mu + (n-c)\xi]P_n + \lambda P_{n-1} + [cq\mu + (n+1-c)\xi]P_{n+1}, c+1 \leq n < N \tag{9}$$
$$0 = -[cq\mu + (N-c)\xi]P_N + \lambda P_{N-1}, n = N \tag{10}$$

The above $(N+1)$ linear equations in the unknown probabilities $P_0, P_1...P_N$ are solved as follows:

Solving (6) and (7), we get

$$P_n = \frac{1}{n!}\left(\frac{\lambda}{q\mu}\right)^n P_0, 0 \leq n \leq c \tag{11}$$

Now, from Eqs. (8)–(10) and using relation (11), we get

$$P_n = \frac{1}{c!}\left(\frac{\lambda}{q\mu}\right)^c \frac{\lambda^{(n-c)}}{\prod_{m=c+1}^{N}[c\mu q + (m-c)\xi]} P_0, c+1 \leq n \leq N \tag{12}$$

Thus, P_n can be written as:

$$P_n = \begin{cases} \frac{1}{n!}\left(\frac{\lambda}{q\mu}\right)^n P_0 & 0 \leq n \leq c \\ \frac{1}{c!}\left(\frac{\lambda}{q\mu}\right)^c \frac{\lambda^{(n-c)}}{\prod_{m=c+1}^{N}[c\mu q + (m-c)\xi]} P_0 & c+1 \leq n \leq N \end{cases} \tag{13}$$

Where P_0 can be obtained using normalization equation, $\Sigma_{n=0}^{N} P_n = 1$.

We use a numerical method (Runge-Kutta method of fourth order) to obtain transient solution of the model. The "ode45" function of MATLAB software is used to compute the transient numerical results. The mean number of tasks waiting in the queue, $L_q(t)$ and the mean waiting time $W_q(t)$ respectively are given by [24]:

$$L_q(t) = \sum_{n=c}^{N}(n-c)P_n(t) \tag{14}$$

$$W_q(t) = \frac{L_q(t)}{c\mu[1 - \sum_{n=0}^{c} P_n(t)]}$$

The transient state probabilities, including the probability that the queue is empty and the probability that the buffer is full can be obtained by solving set of equations in (5) numerically. If the queue is empty, incoming tasks will be processed immediately, it provides good quality of service (QoS) to the users but it is not profitable for the CSPs. If the buffer is full, then incoming tasks will be rejected, which results in poor QoS.

4   Numerical Examples

In this section we present numerical examples to illustrate the effect of reneging and feedback on the queueing delay, probability of task rejection, and probability of immediate service. We use a numerical method (Runge-Kutta method of fourth order) to obtain transient solution of the model. The "ode45" function of MATLAB software is used to compute the transient numerical results.

Figures 3 and 4 shows the variation of the state probabilities with time. Generally, the state probabilities increase sharply and then attains steady state. $P_0(t)$ is the probability that the queue is empty at the time, t, such that any packet that arrives is immediately scheduled into the VM for processing while $P_{10}(t)$ is the probability that there are 10 tasks in the queue. The values of the parameters are taken as: $\lambda = 12, \mu = 5, q = 0.9, \xi = 0.4, c = 3$.

Figure 5 shows the transient behaviour of the mean number of tasks in the queue with time. The mean queue size increases with time for a constant arrival rate and then attains a steady state after a long time. It can be observed that when tasks that are dropped from the queue are resubmitted, the queue size is relatively larger. Similar behaviour can be observed in Fig. 5 and 6, which shows the transient behaviour of the mean delay and the probability of task dropping when the buffer is full. It can also be observed that when the tasks that are dropped from the queue are resubmitted, the probability of task dropping or tail dropping of packets when the buffers are full is relatively higher. The values of parameters for used are: $\lambda = 85, \mu = 30, q = 0.85, \xi = 0.4, c = 3, N = 50$ and the initial condition is $P_7(0) = 1$.

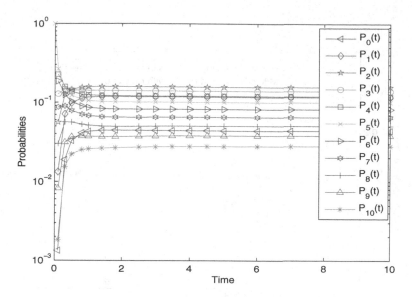

Fig. 3. Probabilities vs time

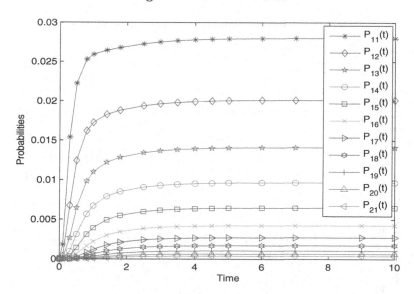

Fig. 4. Probabilities vs time

Figure 7 shows the effect of the average arrival rate of tasks on mean queueing delay. The mean delay increases as the rate of arrival of tasks increases slowly, and after a certain value of the arrival rate, a small increase in the arrival rate of tasks will result in a corresponding fast increase in the delay. A similar behaviour

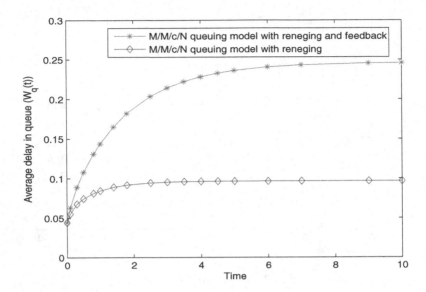

Fig. 5. Comparison of average delay in queue vs time

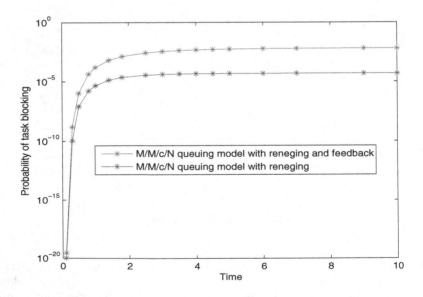

Fig. 6. Comparison of probability of task blocking vs time

of the probability of task blocking can be seen in Fig. 8. The values of parameters are: $\mu = 30, q = 0.85, \xi = 0.3, c = 3, N = 50$ at $t = 3$. Initial condition is $P_7(0) = 1$.

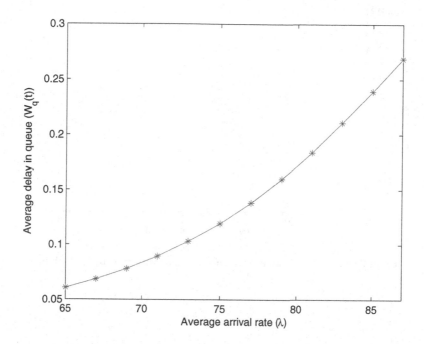

Fig. 7. Effect of average arrival rate on average delay in queue

Figure 9 shows the variation of the mean queueing delay with the probability of feedback. As the probability that tasks that are dropped from the queue are resubmitted increases, the higher the delay. Figure 10 shows that variation of the reneging rate with the mean queueing delay. Resubmission of tasks that reneged from the queue or a task that was rejected is very important to ensure QoS of some users; other users may have to wait longer in the queue. The values of parameters used are: $\lambda = 78, \mu = 30, \xi = 0.3, c = 3, N = 50$ at $t = 3$. Initial condition is $P_7(0) = 1$ and $\lambda = 88, \mu = 30, q = 0.8, c = 3, N = 50$ at $t = 3$. Initial condition is $P_7(0) = 1$ respectively.

Figure 11 shows that increasing the number of VMs will significantly decrease the queueing delays. In other to reduce energy consumption in cloud data centres, but the drawback of such a strategy is an increase in the queueing delay. Similar behaviour can be seen in Fig. 12, where increasing the number of VM also decreases the probability of task blocking. Therefore, increasing the number of VMs will improve the QoS but increases the energy consumption and hence the costs on the CSP. Other QoS and energy optimization methods such as task

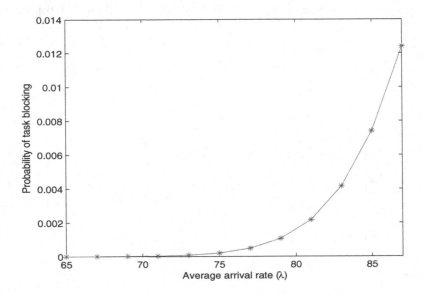

Fig. 8. Effect of average arrival rate on probability of task blocking

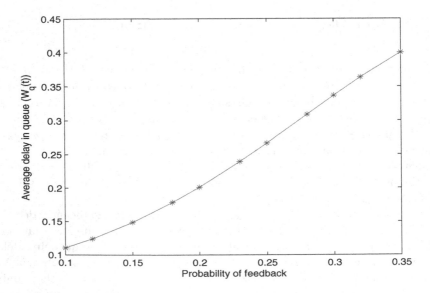

Fig. 9. Effect of probability of feedback on average delay in queue

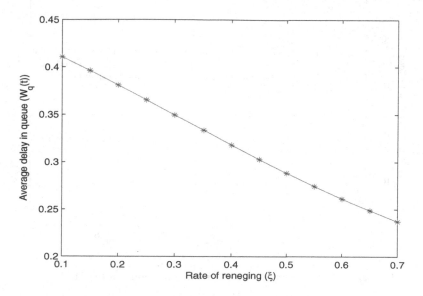

Fig. 10. Effect of rate of reneging on average delay in queue

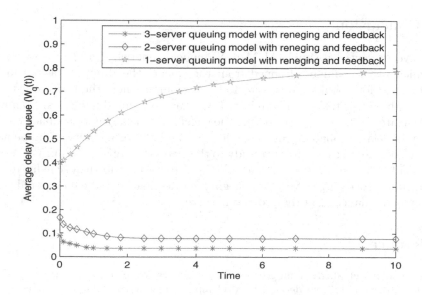

Fig. 11. Effect of number of servers on average delay in queue

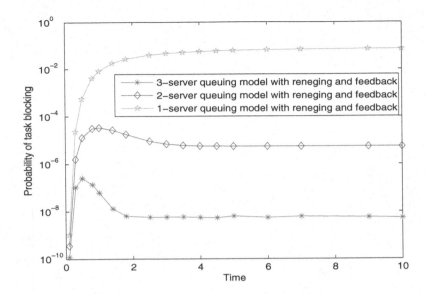

Fig. 12. Effect of number of servers on probability of task blocking

migration can be considered by are out of the scope of this work. The values of parameters used are: $\lambda = 14, \mu = 15, q = 0.8, \xi = 0.1, N = 20$. Initial condition is $P_7(0) = 1$.

5 Conclusion

We have presented a simple M/M/c/N queueing model of a cloud processing in which tasks could be dropped from the queue, the dropped tasks can be resubmitted for possible processing and if the buffer where the tasks are stored is full, subsequent tasks will be rejected. We have presented numerical examples to illustrate its utility by considering the effects of reneging and feedback on the queueing delay, probability of task rejection, and the probability of immediate service. We intend to extend this study to the evaluation of a cloud infrastructure with load balancing, where tasks are the first queue up and then scheduled into the various processing server and reneging and feedback will be considered both at the load balancer and the processing servers.

References

1. Buyya, R., et al.: A manifesto for future generation cloud computing: research directions for the next decade. ACM Comput. Surv. **51**(5), 38 (2018). https://doi.org/10.1145/3241737. Article no. 105
2. Paya, A., Marinescu, D.C.: Energy-aware load balancing and application scaling for the cloud ecosystem. IEEE Trans. Cloud Comput **5**(1), 15–27 (2017)

3. Bruneo, D.: A stochastic model to investigate data center performance and QoS in IaaS cloud computing systems. IEEE Trans. Cloud Comput. **25**(3), 560–569 (2014)
4. Chiang, Y.J., Ouyang, Y.C., Hsu, C.H.: Performance and cost-effectiveness analyses for cloud services based on rejected. IEEE Trans. Serv. Comput. **9**(3), 446–455 (2016)
5. Homsi, S., Liu, S., Chaparro-Baquero, A., Bai, O., Ren, S., Quan, G.: Workload consolidation for cloud data centers with guaranteed QoS using request reneging. IEEE Trans. Parallel Distrib. Syst. **28**(7), 2103–2116 (2017)
6. Ait El Mahjoub, Y., Fourneau, J.-M., Castel-Taleb, H.: Analysis of energy consumption in cloud center with tasks migrations. In: Gaj, P., Sawicki, M., Kwiecień, A. (eds.) CN 2019. CCIS, vol. 1039, pp. 301–315. Springer, Cham (2019). https://doi.org/10.1007/978-3-030-21952-9_23
7. Mishra, S.K., Sahoo, B., Parida, P.P.: Load balancing in cloud computing: a big picture. Advances in Big Data and Cloud Computing. J. King Saud Univ. - Comput. Inf. Sci. (2018)
8. Gupta, S., Arora, S.: Queueing system in cloud services management: a survey. Int. J. Pure Appl. Math. **119**(12), 12741–12753 (2018)
9. Vilaplana, J., et al.: A queueing theory model for cloud computing. J. Supercomput. **69**(1), 492–507 (2014)
10. Czachórski, T., Kuaban, G.S., Nycz, T.: Multichannel diffusion approximation models for the evaluation of multichannel communication networks. In: Vishnevskiy, V.M., Samouylov, K.E., Kozyrev, D.V. (eds.) DCCN 2019. LNCS, vol. 11965, pp. 43–57. Springer, Cham (2019). https://doi.org/10.1007/978-3-030-36614-8_4
11. Vetha, S., Devi, V.: Dynamic resource allocation in cloud using queueing model. J. Ind. Pollut. Control **33**(2), 1547–1554 (2017)
12. Cheng, C., Li, J., Wang, Y.: An energy-saving task scheduling strategy based on vacation queuing theory in cloud computing. Tsinghua Sci. Technol. **20**(1), 28–39 (2015)
13. ElKafhali, S., Salah, K.: Modelling and analysis of performance and consumption in cloud data centers. Arab. J. Sci. Eng. **43**, 7789–7802 (2018)
14. Duan, Q., Yu, S., Zhang, Z.: Cloud service performance evaluation: status, challenges, and opportunities - a survey from the system modeling perspective. Digit. Commun. Netw. **3**, 101–111 (2017)
15. Al-Seedy, R.O., El-Sherbiny, A.A., El-Shehawy, S.A., Ammar, S.I.: Transient solution of the $M/M/c$ queue with balking and reneging: a survey. Comput. Math. Appl. **57**(8), 1280–1285 (2009)
16. Kumar, R., Sharma, S.K.: M/M/1 feedback queueing models with retention of reneged customers and balking. Am. J. Oper. Res. **3**(2A), 1–6 (2013)
17. Karina, V., Rodriguez, Q., Guillemin, F.: Performance analysis of resource pooling for network function virtualization. Psicologia: Reflexao e Crítica, Universidade Federal do Rio Grande do Sul, 2016. hal-01621281 (2016)
18. Chiang, Y., Ouyang, Y.: Profit Optimization in SLA-Aware Cloud Services with a Finite Capacity Queuing Model Mathematical Problems in Engineering. Hindawi Publishing Corporation, London (2014)
19. Farahnakian, F., Pahikkala, T., Liljeberg, P., Plosila, J., Hieu, N.T., Tenhunen, H.: Energy-aware VM consolidation in cloud data centers using utilization prediction model. IEEE Trans. Cloud Comput. **7**(2), 524–536 (2019)
20. Arunarani, A., Manjula, D., Sugumaran, V.: Task scheduling techniques in cloud computing: a literature survey. Future Gener. Comput. Syst. **91**, 407–415 (2019)

21. Abdullahi, M., Ngadi, M.A., Abdulhamid, S.M.: Symbiotic organism search optimization based task scheduling in cloud computing environment. Future Gener. Comput. Syst. **56**, 640650 (2016)
22. Wang, W., Gelenbe, E.: Adaptive dispatching of tasks in the cloud. IEEE Trans. Cloud Comput. **6**(1), 33–45 (2018)
23. Wei, L., Foh, C.H., He, B., Cai, J.: Towards efficient resource allocation for heterogeneous workloads in IaaS clouds. IEEE Trans. Cloud Comput. **6**(1), 264–275 (2018)
24. Kumar, R., Soodan, B.S.: Transient numerical analysis of a queueing model with correlated reneging, balking and feedback. Reliab.: Theory Appl. **14**(4), 46–54 (2019)

Author Index

Printed in the United States
By Bookmasters